Current Topics in Microbiology and Immunology

Volume 375

For further volumes:
http://www.springer.com/series/82

Kenneth Palmer · Yuri Gleba
Editors

Plant Viral Vectors

Responsible Series Editor: Yuri Y. Gleba

Springer

Editors
Kenneth Palmer
James Graham Brown Cancer Center
University of Louisville
Owensboro, KY
USA

Yuri Gleba
Nomad Bioscience GmbH
Icon Genetics GmbH
Halle
Germany

ISSN 0070-217X ISSN 2196-9965 (electronic)
ISBN 978-3-642-40828-1 ISBN 978-3-642-40829-8 (eBook)
DOI 10.1007/978-3-642-40829-8
Springer Heidelberg New York Dordrecht London

Library of Congress Control Number: 2013948363

Printed on acid-free paper

Springer is part of Springer Science+Business Media (www.springer.com)

Preface

Plant virologists are fond of reminding their non-plant virologist colleagues of the myriad of contributions that Tobacco mosaic virus has played in the development of modern molecular biology. Our friends and colleagues who organized the 1998 Symposium celebrating 100 years of TMV research *Tobacco mosaic virus*, sponsored by the Royal Society of Edinburgh in Association with the Royal Society, provide a wonderful volume of personal accounts of the history TMV research in the *Philosophical Transactions of the Royal Society of London (Series B) vol. 354*. The contributors to that volume were among those who laid the foundations for the industrialization of plant virology through manipulation of plant virus genetic sequences in design of plant viral vector systems. In this volume, our co-authors provide a wonderful overview of how far the plant viral vector field has come in the intervening years. Our discipline is no longer exclusively in the domain of academics—there is a small, but growing number of small (but growing) biotechnology companies that exploit plant viruses as the platform for commercial innovation in crop improvement, industrial product manufacturing, and human and veterinary healthcare.

Many of us in the plant viral "vectorology" field can trace our scientific pedigree back in some fashion to Bill Dawson, and we are honored that he contributed the opening chapter of this volume with a narrative on the history of plant virus vector development. Bill reminds us that "a vector is not a virus; it is a device designed to perform a specific function." Other contributors have provided fascinating reviews of how plant viral vectors have been adapted to serve specific functions, from plant gene function discovery through nanotechnology and providing infinitely scaleable manufacturing systems for valuable human therapeutics. We thank all of our co-authors for their contributions.

This year marked the death of an esteemed polymath, virologist, vaccinologist Dr. Hilary Koprowski, who late in his illustrious career championed plant viral vectors for their application in solving particularly difficult public health problems. We dedicate this volume to his memory.

Kenneth Palmer
Yuri Gleba

Contents

Contributors

William O. Dawson Department of Plant Pathology, Citrus Research and Education Center, University of Florida, 700 Experiment Station Road, Lake Alfred, FL 33850, USA, e-mail: wodtmv@crec.ifas.ufl.edu

Denis Leclerc Infectious Diseases Research Centre, Pavilion CHUL, Laval University, 2705 Boulevard Laurier, Quebec City, PQ G1V 4G2, Canada, e-mail: Denis.Leclerc@crchul.ulaval.ca

George P. Lomonossoff Department of Biological Chemistry, John Innes Centre, Norwich Research Park, Norwich NR4 7UH, UK, e-mail: george.lomonossoff@jic.ac.uk

David J. Evans Department of Biological Chemistry, John Innes Centre, Norwich Research Park, Norwich NR4 7UH, UK, e-mail: dave.evans@jic.ac.uk

Edward P. Rybicki Institute of Infectious Disease and Molecular Medicine, Cape Town, South Africa; Department of Molecular and Cell Biology, University of Cape Town, PB Rondebosch, Cape Town, Western Cape 7701, South Africa, e-mail: ed.rybicki@uct.ac.za

Darrin P. Martin Institute of Infectious Disease and Molecular Medicine, Cape Town, South Africa

Christophe Lacomme Virology and Zoology Section, Science and Advice for Scottish Agriculture (SASA), Roddinglaw Road, Edinburgh, EH12 9FJ, UK, e-mail: Christophe.Lacomme@sasa.gsi.gov.uk

Victor Klimyuk Icon Genetics, Weinbergweg 22, 06120 Halle (Saale), Germany, e-mail: klimyuk@icongenetics.de

Gregory Pogue Kentucky BioProcessing, LLC, Owensboro, KY, USA; The IC2 Institute, University of Texas at Austin, Austin, TX, USA

Stefan Herz Icon Genetics, Weinbergweg 22, 06120 Halle (Saale), Germany

John Butler Bayer Innovation GmbH, Merowinger Platz 1, 40225 Düsseldorf, Germany

Hugh Haydon Kentucky BioProcessing, LLC, Owensboro, KY, USA

Kevin J. Whaley Mapp Biopharmaceutical Inc, 6160 Lusk Blvd, Suite C105, San Diego, CA 92121, USA, e-mail: Kevin.whaley@mappbio.com

Josh Morton Kentucky BioProcessing, LLC 3700 Airpark Drive, Owensboro, KY 42301, USA, e-mail: JMorton@kbpllc.com

Steve Hume Kentucky BioProcessing, LLC 3700 Airpark Drive, Owensboro, KY 42301, USA, e-mail: SHume@kbpllc.com

Ernie Hiatt Kentucky BioProcessing, LLC 3700 Airpark Drive, Owensboro, KY 42301, USA, e-mail: EHiatt@kbpllc.com

Barry Bratcher Kentucky BioProcessing, LLC 3700 Airpark Drive, Owensboro, KY 42301, USA, e-mail: BBratcher@kbpllc.com

Victor Klimyuk Icon Genetics GmBH, Biozentrum Halle, Weinbergweg 22, Halle (Saale), D-06120, Germany, e-mail: Klimyuk@icongenetics.de

Andrew Hiatt Mapp Biopharmaceutical Inc, 6160 Lusk Blvd, Suite C105, San Diego, CA 92121, USA, e-mail: Andy.hiatt@mappbio.com

Michael Pauly Mapp Biopharmaceutical Inc, 6160 Lusk Blvd, Suite C105, San Diego, CA 92121, USA, e-mail: Michael.pauly@mappbio.com

Larry Zeitlin Mapp Biopharmaceutical Inc, 6160 Lusk Blvd, Suite C105, San Diego, CA 92121, USA, e-mail: Larry.zeitlin@mappbio.com

Yuri Y. Gleba Nomad Bioscience GmbH, Weinbergweg 22, Halle (Saale), Germany, e-mail: gleba@nomadbioscience.com

Daniel Tusé Nomad Bioscience GmbH, Weinbergweg 22, Halle (Saale), Germany

Anatoli Giritch Nomad Bioscience GmbH, Weinbergweg 22, Halle (Saale), Germany

A Personal History of Virus-Based Vector Construction

William O. Dawson

Abstract The ability to express foreign genes or to silence endogenous genes in plants has revolutionized both basic and applied plant biology. Virus-based expression systems, in which the foreign mRNA is greatly amplified by virus replication, can produce very high levels of proteins or peptides in leaves and other tissues. Vectors have been available for about 25 years. They are commonplace as laboratory tools, but their initial commercial expectations have not been met for numerous reasons. Yet, applications of viral vectors are still evolving. This chapter focuses on our laboratory's involvement in developing virus-based vectors in plants. We created the first 'add-a-gene' vectors that were capable of replication and movement throughout plants. These vectors were based on tobacco mosaic virus. Through the evolution of several prototypes, stable vectors were developed that produced relatively large amounts of product in plants. Recently, we created similar vectors for citrus trees based on citrus tristeza virus. Even though the citrus vectors were created as laboratory tools for improving the crop, circumstances have changed the applications to protection and therapy of trees in the field.

"When what needs to be learned changes quickly, especially in the course of a single generation, it becomes harder to know what to teach and how to teach it... In a world in transition, students and teachers both need to teach themselves one essential skill—learning how to learn."Carl Sagan, *A Demon-Haunted World*

W. O. Dawson (✉)
Department of Plant Pathology, Citrus Research and Education Center,
University of Florida, 700 Experiment Station Road,
Lake Alfred, FL 33850, USA
e-mail: wodtmv@crec.ifas.ufl.edu

Current Topics in Microbiology and Immunology (2011) 375: 1–18
DOI: 10.1007/82_2011_182

Published Online: 25 October 2011

Contents

The ability to express foreign genes or to silence endogenous genes in plants has revolutionized both basic and applied plant biology. Foreign genes can be expressed in plants either by permanent insertion into the genome or by transient expression using virus-based vectors. Each approach has distinct advantages. While the insertion of genes into the plant genome is generally permanent, expression by viral vectors occurs only for a limited period of time. In genetically improving plants, these systems are complementary. The potential value and applications of each of these approaches varies broadly, especially comparing annual to woody plants.

This chapter focuses on our laboratory's involvement in developing virus-based vectors in plants. These primitive vectors essentially function as viruses with extra genes. Other laboratories have deconstructed the viruses and utilized their properties to produce vectors that are not viruses. Other chapters will describe different types of vectors and their applications.

Many plant viruses have small genomes that can be manipulated easily. Some of these viruses accumulate to copious amounts in infected plants in relatively short periods of time. The signature example is *Tobacco mosaic virus* (TMV), which can pack virions into almost all available space within tobacco cells (Fig. 1). Virus-based expression systems, in which the foreign mRNA is greatly amplified by virus replication, can produce very high levels of certain proteins in leaves and other tissues. Contrarily, transgenic expression of foreign genes in the plant rarely produces similar levels. Hence virus-based expression is useful in examining the results of over-expression of genes to identify their functions as well as in producing large amounts of a protein. Some foreign proteins of interest are inherently phytotoxic and interfere with growth and development of the transgenic plant. The viral vector systems offer an approach to overcome toxic effects on regeneration and development by directly inoculating mature plants, which are past the stages of regeneration and development. An important feature of the virus vector system is the rapid accumulation of foreign gene products, which can be harvested within one to two weeks after building constructs and inoculating plants, allowing for rapid evaluation of protein function. Another advantage, somewhat ironically, is their instability. Virus vectors always lose inserted sequences such that foreign genes cannot permanently persist in the environment. The desired compromise is that vectors need to be sufficiently stable

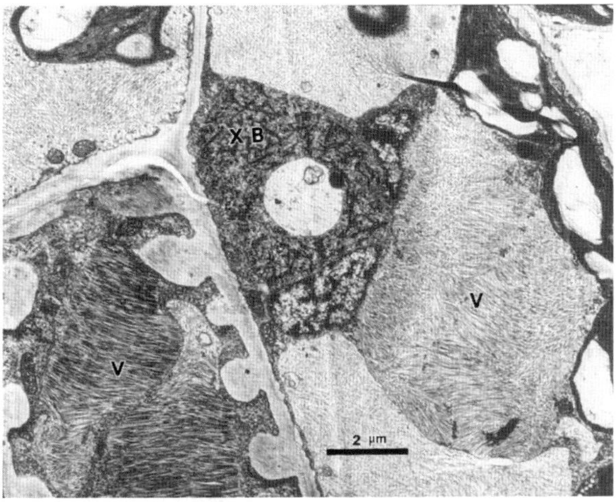

Fig. 1 Electron micrograph of TMV infected cell. *V* is virions; *XB* is X body

to produce products in plants but sufficiently unstable such that the foreign sequence will not remain in the environment for long periods after production. Manipulations affecting recombination, such as the amount and degree of repeated sequences, can be used to adjust stability.

The uses for virus-based vectors seem limited only by our imagination. Originally, researchers sought: (1) to create laboratory tools to mark viruses for visualization of virus movement and distribution in plants; (2) to examine the effect of foreign gene expression in plants; and (3) to remove expression of plant genes by silencing to examine the resulting effects on plant phenotypes. The idea of producing commercial products followed soon afterwards, as researchers speculated on the potential of growing some specialty products in natural acreage, while other more valuable products could be produced in a greenhouse or laboratory setting. Numerous commercial products have been considered from small amounts of high value pharmaceutical products to large amounts of crude preparations of enzymes to be used in industrial processes. Considerable efforts currently are being made to use vectors to produce immunogens or antibodies for disease control and diagnosis. The rapid turn-around from construction to product has been utilized for production of individualized treatment of human patients against certain diseases. Also, development of high-through-put systems of utilizing vectors has been used as a genomics tool for identifying gene functions. Recently, our laboratory began considering a new use of virus-based vectors—as therapeutic vehicles for treating trees. As I perceive the evolution of this work, various uses for virus vectors fit nicely into what always has been my definitions of basic versus applied research: Applied research occurs in situations when there is an identified problem but an unidentified solution; basic research occurs in situations when there is an identified solution but an unidentified problem. Virus-based vectors are the solution; our challenge is to identify appropriate problems.

To date, virus-based vectors have predominantly been used on annual, herbaceous plants. However, it is possible that more valuable uses of these vectors would be in perennial woody plants. Certainly, genetically improving tree crops, because of their complex breeding and necessary time to reach maturity, offers unique challenges to the researcher. Transformation of woody trees can take up to a year to produce a small plantlet, and 5–20 years to examine the effect of the transgene on mature tree characteristics—particularly traits such as yield potential and fruit or nut quality. The challenges for improving trees greatly increase the need of useful virus-based vectors. Varietal improvement through genetic engineering would benefit tremendously from the time-savings of an effective vector that could enable rapid assays for potential gene insertions or knock-outs. Transient expression of potential genes in mature trees would allow the evaluation of numerous genes in a relatively short period of time, after which the selected genes would be transformed into the desired tree line. Thus, the transient-expression vector can be a valuable tool to complement stable transformation.

Another value of the virus-based vector is the ease of moving the gene to other trees. Once it infects a tree, a vector can be transferred to unlimited numbers of other trees of different varieties or species or different ages, including mature trees by mechanical or graft inoculation. The vector could be used to quickly screen genes with potential to improve trees, evaluated at different growth stages, and evaluated in different varieties.

However, the use of virus-based vectors in trees might require development of more stable vectors than those of the first generations used in herbaceous plants, because most assays in trees require longer periods than those in herbaceous plants. If more stable vectors can be developed, a potential use in perennial crops that has not been explored yet is to treat existing plants in the field. The inherent high value of individual trees increases the need and value of effective transient vectors. For example, foreign genes could be added to trees to protect them from pests or diseases, or even to treat them after they were affected to overcome the disease or pest. Since tree crops are capable of production for many years, it is common for emerging threats to appear long after the trees reach maximal production. Removing the trees and replacing them with another generation of trees that are resistant or tolerant to the threat can be extremely costly. Vectors could be designed to protect them from new threats as they emerge.

1 Evolution of Virus-Based Vectors

Virus-based vectors have been built empirically, evolving in small steps as we have learned more about regulation of viral replication and gene expression. In general, the best vectors have resulted from luck based on what a particular virus allowed. Vector development resulted from the development of recombinant DNA technologies, which emerged in the 1980s, and provided tools that allowed virologists to do 'reverse genetics' for the first time. The subsequent development of infectious clones of plant viruses provided opportunities to play new games. One such game rearranged viral genomes by removing genes, inserting genes, and/or misplacing genes to see what would happen. One obvious scenario inserted foreign genes, or more correctly foreign open reading frames (ORFs), into the virus to see if they were expressed—thus, the virus-based expression vector.

However, as these genetic systems were being developed, controversy arose concerning the future of virus-based vectors. Al Siegel was an early visionary in the field advocating the value of virus-based transient expression vectors (Siegel 1983). However, Lois Vloten-Doting et al.—along with many other people in the field—argued that because of the low fidelity of RNA virus replication and the lack of a proof-reading mechanism, RNA viruses would have little value as vectors because only viral genes under selection could be maintained—foreign sequences would quickly drift to randomness (van Vloten-Doting et al. 1985). Yet, when constructs were built to test this concern, we found the rate of mutation accumulation in foreign sequences of a TMV-based vector to be sufficiently low that sequence drift was not a practical limitation to the use of TMV vectors (Kearney et al. 1993).

The first attempts at producing vectors from RNA viruses had a foreign ORF substituted for the viral ORF for the coat protein (for example S3-CAT-28 of Fig. 2) (French et al. 1986; Takamatsu et al. 1987; Dawson et al. 1988; Chapman et al. 1992; Scholthof 1999). With viruses like *Brome mosaic virus* (BMV), which need the coat protein for cell to cell movement, resulting vectors were essentially limited to protoplasts. They were unable to move from initially inoculated cells in plants. Other vectors based on viruses like TMV, which did not need the coat protein to move cell to cell, could move within inoculated leaves, but were unable to move systemically into upper leaves. None of these vectors held much utility.

2 First Generation of Add-a-Gene Vectors

Our laboratory was determined to create a vector that would express foreign genes systemically throughout plants. Because of its extremely high level of protein production, TMV was chosen as the basis of such a vector. TMV is a positive-sense RNA virus of 6,395 nucleotides, expressing two replicase-associated proteins from the genomic RNA, and the movement and coat proteins from 3′ coterminal subgenomic mRNAs (Dawson and Lehto 1990). The level of production of the TMV coat protein, which accumulates to as much as 10% of the dry weight of an infected leaf, could be considered a goal for an optimal virus-based expression of a foreign protein. We attempted to build a vector that would contain all of the viral genes needed for replication and movement by adding the foreign gene as an extra gene in TMV (Dawson et al. 1989). Initially, the foreign ORF of chloramphenicol acetyltransferase (CAT) was engineered to be controlled by an additional coat protein subgenomic RNA promoter into TMV either before (CAT-CP) or after (CP-CAT) the coat protein gene (Fig. 2). However, these first attempts were disappointing. CP-CAT moved very poorly and CAT-CP was too unstable. The foreign ORF of CAT-CP was surrounded by the repeated sequences of the coat protein promoter. Homologous recombination between those sequences resulted in deletion of the inserted sequences and the recombinant was wild-type TMV. Recombination occurred so quickly that it was difficult to retain and assay the vector (Dawson et al. 1989).

3 Second Generation Add-a-Gene Vectors

Meanwhile, other researchers in our laboratory were manipulating TMV in different ways, providing leads for building the next prototypical vector.

- Kirsi Lehto was examining the movement protein gene by examining TMV constructs with: (1) two movement proteins; (2) the movement protein in different locations; and (3) the movement protein controlled by the coat protein subgenomic RNA promoter (Lehto and Dawson 1990a,b; Lehto et al. 1990). Remarkably, she found that the construct with the coat protein promoter controlling the movement protein was stable even though the coat protein promoter sequences were repeated on each side of the movement protein ORF, the same as the previous unstable vector. We realized that the stability was due to the fact that homologous recombination in this construct resulted in a virus that lost the movement gene, and thus, was selected against because it was not capable of spreading within the plant.
- David Beck was examining expression of the coat protein gene in TMV (Beck and Dawson 1990). He constructed a series of viruses with two mutant copies of the TMV coat protein. These were the least stable constructs that we examined. The two coat proteins were never detected in infected plants. Here, homologous recombination quickly resulted in a virus with only one coat protein gene.
- Mark Hilf built a similar virus with two coat protein genes, one on from TMV and the other from a different tobamovirus, *Odontoglossum ringspot virus* (ORSV). In contrast, this two-coat-protein virus was stable (Hilf and Dawson 1993; Culver et al. 1993).

These data suggested that recombination was fostered by exact repeats of sequences in TMV and that avoidance resulted in substantial increases in stability. It is rare to know that the next proposed experiment is going to work. However, these results led to the design of a second generation of an add-a-gene vector, which used a subgenomic RNA promoter from a different tobamovirus as the extra promoter. This vector, referred to as TB2, used the TMV coat protein promoter to control the foreign ORF (Donson et al. 1991). The coat protein was supplied by the promoter and ORF from ORSV (Fig. 2). This vector was relatively stable, expressing the foreign gene throughout infected plants. TB2 was sufficiently stable to produce useful amounts of foreign proteins in plants (Donson et al. 1991; Kearney et al. 1993). Moreover, it produced the biologically active a-trichosanthin,

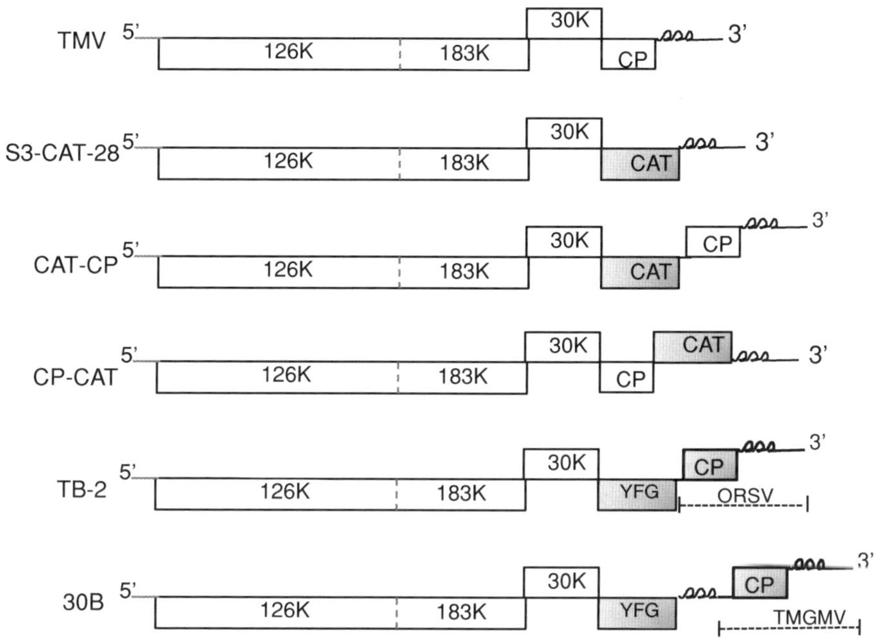

Fig. 2 Schematic diagram of the genome organization of the wild type TMV and TMV-based transient expression vectors. The open boxes represent ORFs and their translation products. *30 K* is the movement protein; *CP* is the coat proteins; *CAT* is chloramphenicol acetyltranferase; *YFG* is your favorite gene; *ORSV* is *Odontoglossum ringspot virus*; *TMGMV* is *Tomato mild green mosiac virus*. Inserted elements are shown in *grey* and lines are *darkened*

a potential therapeutic in the treatment of AIDS (Kumagai et al. 1993). It was also shown that a similar vector construct could be used to silence host genes (Kumagai et al. 1995).

4 Third Generation Add-a-Gene Vectors

The level of foreign gene expression regulated by the TMV coat protein subgenomic (sg) mRNA promoter in TB2 resulted in moderate levels of foreign protein production, but much lower than the levels of coat protein produced by wild-type TMV or even the heterologous coat protein of the vector. We were determined to develop a higher yielding vector. TB2 was arbitrarily built from two different tobamoviruses, TMV and ORSV. However, a number of other tobamoviruses with different levels of sequence similarity had potential for building vectors that might work better. Since it was not known how different tobamovirus sequences might affect the performance of TMV-based expression vectors, a series of chimeric vectors based on different tobamovirus sequences were examined (Shivprasad et al. 1999).

Meanwhile, an increased understanding of what affects tobamovirus gene expression was being developed. For example, levels of expression of internal genes in TMV depended on the proximity of the ORF to the 3′ terminus (Culver et al. 1993). The highest expressed TMV gene is that of the coat protein, which is adjacent to the 3′ NTR. The 3′ NTR is a complex structure that consists of three pseudoknots followed by the tRNA-like structure (Pleij et al. 1987). We found the movement protein gene, which in its native position is internal to the coat protein gene, to be expressed in much smaller amounts, but that levels of production of the movement protein could be greatly altered by changing its relative position of the gene within the viral genome. When the movement protein gene was placed closer to the 3′ terminus by deleting part of or all of the coat protein gene, its synthesis was progressively increased up to 20 times that of the wild-type virus (Culver et al. 1993). Insertions that moved the movement protein gene further from the 3′ terminus correspondingly decreased protein production. Position also affected the expression of the coat protein gene. In a virus with two coat protein genes in tandem, the level of expression of the 3′ proximal gene was 10–100 times more than the more 5′ gene (Culver et al. 1993). Moreover, insertion of the three pseudoknots from the TMV 3′ NTR between the two coat protein genes caused the level of expression of the 5′-most gene to increase to levels similar to that of the 3′-most gene (Chapman et al. unpublished results), suggesting that proximity of the gene to the pseudoknots was a key to increased expression. Based on this information, a series of vectors were designed with a duplication of the three pseudoknots of the TMV 3′ NTR internally between the foreign ORF and the coat protein gene (Shivprasad et al. 1999).

The different hybrid vectors expressing GFP were essentially from the same design, but containing different heterologous tobamovirus sequences including an extra subgenomic mRNA promoter. Yet, there were substantial differences in the amounts of GFP produced by the different vectors, clearly demonstrating that building an effective vector from a virus is not a trivial exercise (Shivprasad et al. 1999). The performance of the vector was based on a set of rules most of which were not understood by us. However, one concept that became evident was the competition for production of sg mRNAs of the different genes. Apparently there is a limited capacity for production of sg mRNAs by the TMV replication complex. Thus, an effective expression vector should appropriately partition the limited replicase activity among the various promoters to ensure adequate replication and movement of the vector while providing the maximal level of foreign gene expression. Even though there is a preference for increased transcription of subgenomic mRNA promoters nearer the 3′ terminus, this could be countered by using promoters that are less efficient. Also, insertion of an extra set of pseudoknots between the foreign gene and the heterologous coat protein gene increased expression of the internal foreign gene.

The critical measurement for a virus-based expression vector is the level of protein production. The vector 30B (Fig. 2) was the winner from the series of different hybrid vectors (Shivprasad et al. 1999). It had reduced coat protein sgRNA production, which allowed more foreign gene sgRNA production, but it produced enough coat protein for the virus to spread throughout plants. This vector quickly invades and produces large amounts of protein throughout *Nicotiana benthamiana* plants. The foreign protein generally was produced to levels that were readily visible bands in stained polyacrylamide gels. However, with foreign ORFs the size of GFP or larger, 30B does not move efficiently through phloem in tobacco, perhaps due to some genome length constraint. Yet, it effectively produces foreign proteins in inoculated leaves of tobacco plants. One important characteristic of virus-based vectors is the degree of stability. The vector needs to be sufficiently stable to produce products in plants, but it must also be sufficiently unstable so the foreign sequence will not remain in the environment long after production. 30B could be propagated in *N. benthamiana*, allowing full systemic infection over a 2–3 week period. Inoculum from the upper leaves could be used to infect a new set of plants resulting in a similar systemic infection with uniform fluorescence. This could be repeated through three or four serial passages in plants.

5 Citrus Tristeza Virus-Based Expression Vectors

The next vectors we created were for citrus trees based on *Citrus tristeza virus* (CTV). Interestingly, the value and potential uses of these vectors have changed continually as different threats and new ideas have emerged.

CTV is a member of the complex *Closteroviridae* family that contains viruses with mono-, bi-, and tri-partite genomes (Bar-Joseph et al. 1979; Agranovsky 1996; Karasev 2000; Dolja et al. 2006). The long flexuous virions (2000 nm × 10–12 nm) of CTV are encapsidated by two coat proteins, the major coat protein (CP) covering about 97% of the virion and the minor coat protein (CPm) completing encapsidation of the 5′ end of the RNA. The single-stranded RNA genome of CTV is 19.3 kb, divided into twelve open reading frames (ORFs) (Pappu et al. 1994; Karasev et al. 1995) (Fig. 3). ORF 1a encodes a 349 kDa polyprotein containing two papain-like protease domains plus methyltransferase-like and helicase-like domains. Translation of the polyprotein occasionally continues through the polymerase-like domain (ORF 1b) by a +1 frameshift. ORFs 1a and 1b plus the non-translated termini are all that is required for replication in protoplasts. Ten 3′ ORFs are expressed by 3′ co-terminal subgenomic (sg) mRNAs (Hilf et al. 1995; Karasev et al. 1997). These 3′ ten genes are involved in movement and interactions with the hosts. In addition to the two coat proteins, p65 (hsp70 homologue) and p61 are required for efficient virion assembly, and are necessary for passage of the virus from protoplast to protoplast in order to amplify inoculum for infection of citrus trees (Satyanarayana et al. 2000). The p6 protein is needed for infection of plants as are the p20 and p23 proteins, which along with CP are suppressors of RNA silencing (Lu et al. 2004).

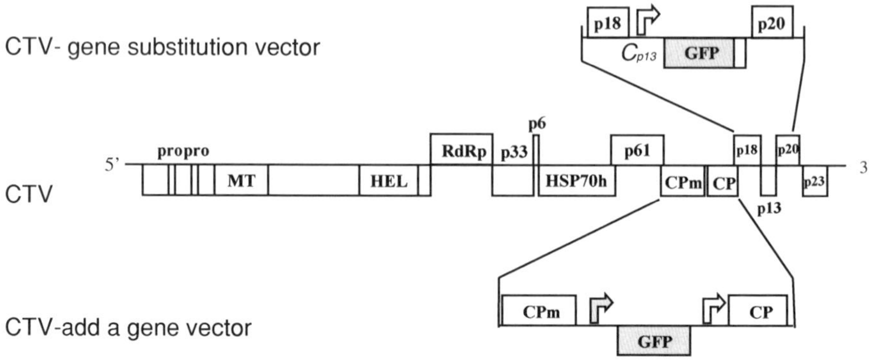

Fig. 3 Schematic diagram of the genome organization of the wild type CTV and CTV-based transient expression vectors. The open boxes represent ORFs and their translation products. *PRO*-papain-like protease domain; *MT*-methyltransferase; *HEL*-helicase; *RdRp*-RNA-dependent RNA polymerase; *HSP70 h*-HSP70 homolog (p65); *CPm*-minor coat protein; CP-major coat protein; *GFP*-green fluorescent protein. *Bent arrows* indicate positions of sgRNA controller elements. Inserted elements are shown in *grey*

There appear to be general rules that determine the regulation of the 3′ genes. The first rule maintains that genes located nearer the 3′ terminus are usually expressed at higher levels. The two 3′-most genes, p23 and p20, have the highest levels of sgRNAs. Also, when genes are moved closer to the 3′ terminus, expression levels increase. Positioning the lowly expressed p33 gene near the 3′ terminus results in a level of expression comparable to the highest expressed genes (p20 and p23) (Satyanarayana et al. 1999). However, as with many rules, there are exceptions. The CP gene, located at position 5 from the 3′ end, is expressed higher than the p13 and p18 genes, located at positions 3 and 4. The second rule maintains that ORFs with an upstream non-translated region are generally expressed higher than ORFs that overlap with the preceding ORF. With the exception of the 5′—most genes (p33 and p6), these are the more highly expressed genes. The cis-acting sequences (which we refer to as controller elements) that regulate the expression of the 3′ genes generally are located immediately upstream of its ORF (Gowda et al. 2001, 2003; Ayllón et al. 2003, 2004, 2005).

An infectious cDNA clone of CTV serves as the basis of our reverse genetics system and allows replication of the recombinant virus in protoplasts (Satyanarayana et al. 1999). However, a complete genetic system requires that the cycle be completed from the infected tree through cDNA and back to a tree infected with the recombinant virus. Although we have had no success in directly inoculating citrus trees from the cDNA, using either RNA transcripts or *Agrobacterium*-inoculation, we were successful in inoculating small citrus trees with virus that was amplified through successive cycles in protoplasts (Satyanarayana et al. 2001).

The earliest prototype plant virus-based vectors were gene-substitution vectors. These vectors, in which the coat-protein gene was replaced, did not systematically infect plants; however, CTV has potential to function as a gene-substitution vector because it has three genes—p33, p18, and p13—that can be deleted with continued efficient replication and movement of the virus in some citrus varieties (Tatineni et al. 2008). The p33 and p18 genes are expressed at relatively low levels, but the p13 gene is expressed at moderate levels. Thus, we chose to replace the p13 ORF with the ORF of GFP (Fig. 3). However, the level of expression of GPF by this construct was disappointing (Folimonov et al. 2007). Yet, this failure of substituting a foreign ORF for the p13 ORF does not eliminate this strategy as viable. It is possible that we destroyed necessary cis-acting elements within the p13 ORF of which we were unaware. This only emphasizes our inadequate understanding of regulation of viral gene expression. Expression of a foreign gene from this position is still attractive, particularly for a second or third foreign gene from a CTV-based vector. Positioning the foreign ORF differently behind a large segment of sequences containing the controller element or the use of a stronger controller element might allow effective expression from the ORF replacement.

The 'add-a-gene' vector, with an extra controller element plus the GFP ORF inserted between the coat protein genes at position 6 from the 3′ end (Fig. 3), worked well (Folimonov et al. 2007). Although CTV is limited to phloem-associated cells, the levels of foreign gene production by the CTV based vectors in infected cells appeared to be similar to that in cells infected with the TMV-based vectors. This vector has been very useful in allowing visualization of CTV movement and distribution in citrus (Folimonova et al. 2008) and the interactions between different strains of CTV (Folimonova et al. 2010). The foreign gene position, between the two coat proteins genes, was chosen arbitrarily. Since genes located nearer the 3′ terminus tend to be expressed at higher levels, it appears that more 3′ positions would work even better. In fact, we have found that CTV effectively expresses high levels of foreign genes from several other locations within its genome (El Mohtar and Dawson, unpublished).

The regulation of the sgRNAs of the closterovirus is completely different from that of the tobamoviruses. In examination of an extra gene in TMV, it was clear that there was competition between the different sgRNAs—increases in one sgRNA resulted in decreases in others (Shivprasad et al. 1999). In contrast, there appeared to be no competition in production of the different sg mRNAs of the CTV-based vectors (Folimonov et al. 2007). Inserting a new gene or changing the levels of expression of different genes had little or no effect on the levels of subgenomic mRNA production by the other genes (Ayllón et al. 2004; Folimonov et al. 2007).

A limitation of virus-based vectors has been their relatively low stability. For example, the vectors used most widely as laboratory tools are based on TMV and PVX, but their progeny populations are overcome by recombinants that have lost most of the inserted sequences during the systemic infection of their hosts within weeks. In contrast, several CTV-based vectors have remained the major component of the population for years. The originally infected plants have continually expressed GFP for more than seven years. In fact, the uniqueness of the CTV-based vectors is their stability. Instability generally has been blamed on high error rates of RNA virus replication and high rates of recombination. Yet, this virus in nature appears to be unusually stable. Sequences of different isolates of CTV maintained in different countries and in different varieties of citrus for more than a hundred years are essentially identical (Albiach-Martí et al. 2000), and the progeny of cloned CTV did not appreciably change during six years of propagation (Xiong et al. unpublished). However, recombination of CTV does not appear to be limited. Most wild populations contain many defective RNAs that apparently resulted from facile recombination (Mawassi et al. 1995). For TMV-based vectors, it was shown that repeated sequences decreased stability, apparently by increasing recombination rates (Dawson et al. 1989; Donson et al. 1991). This instability could be partially overcome by avoiding exact repeated sequences by using promoters from different tobamoviruses as the extra promoter (Donson et al. 1991; Rabindran and Dawson 2001). However, repeated sequences in the CTV-based vector resulting from duplicated CP promoter elements did not appreciably decrease stability (Folimonov et al. 2007). The stability of the CTV-based vectors appears not to be caused by reduced recombination, but instead by increased competitiveness with the potential wild-type-like recombinant. When inoculated simultaneously with the wild-type virus, the CTV-based vectors with an extra gene were able to compete effectively with the wild-type virus during replication and movement throughout citrus trees. Even though recombinants that have the inserted sequences deleted are produced in the CTV populations, there appears to be little selection for these recombinants to be increased proportionally in the populations. In contrast, the TMV-based vectors compete poorly with the wild-type-like recombinants for both cell to cell and long-distance movement (Rabindran and Dawson 2001). Thus, the TMV recombinants quickly overcome the vector during spread in the plant.

CTV is a large virus with ten 3′ genes expressed through sg mRNAs. The addition of an extra gene had no obvious effect on the virus. The number of additional genes the virus will tolerate is unknown, but it appears likely to tolerate more than one. The virus can tolerate deletion of three genes with little apparent effect on its ability to replicate, move, and express the other genes in citrus trees. It is possible that each of these ORFs could be substituted by a foreign ORF. Also, extra genes can be inserted between CTV genes at numerous locations. Thus, it is possible that a vector could be designed to express multiple foreign genes. In fact, we now have vectors expressing three reporter genes (El Mohtar and Dawson, unpublished).

CTV, and thus CTV-based vectors, are generally limited to phloem-associated cells. Phloem-limitation of a vector would be expected to produce less foreign gene product in plants than vectors that infect most cells of the plant. This limitation can be overcome in some situations. For example, we produced antimicrobial peptides against the citrus canker bacterium, which is intercellular, by engineering leader sequences to export the peptides into the intercellular space where they were dispersed in the liquid films between cells. Some of the antimicrobial peptides tested were able to limit the multiplication and spread of the canker bacterium in citrus. Yet, for some uses, the phloem limitation can be an advantage. As discussed in the next section, the CTV vector can be used to treat

Fig. 4 Two pictures of the same row of citrus before and after becoming severely affected by HLB. Photographs courtesy of Mike Irey

citrus trees infected with citrus greening because both the virus and bacterium reside in the phloem.

6 How Circumstances Changed the Value of the CTV-Based Vector

Huanglongbing (HLB, also known as citrus greening) is the limiting factor in citrus production in many countries throughout the world. The disease is caused by the α-protobacterium, *Candidatus* Liberibacter asiaticus (Las), which resides in the phloem of infected citrus trees. The bacterium also multiplies in the Asian citrus psyllid (ACP, *Diaphorina citri*), which transmits the disease from tree to tree. The disease was discovered in Florida in 2005 and has spread quickly throughout the state. Figure 4 shows the effect the disease is having on citrus trees. Presently, the only control methods for HLB are reducing psyllid populations and quickly identifying and removing infected trees. However, this approach is so costly in terms of applying insecticides and removing trees that it is rapidly being abandoned.

In the long term, the most economical and sustainable approach to managing the citrus greening disease will be the production of resistant trees. Citrus, particularly sweet orange and grapefruit, is not amenable to traditional breeding, especially in a short period of time. Resistant trees will likely have to result from transgenes. Initially the CTV vector was used to screen potential genes that would resist Las or repel the psyllid, with the plan to insert those genes into transgenic citrus. However, it will likely take 15 years for transgenic citrus to be evaluated and approved for commercial use, and another 15 years to produce enough new trees to replace existing groves. HLB has spread much faster than expected and in combination with other dilemmas such as loss of groves to hurricanes, suburban sprawl, and a canker eradication program, it is questionable whether the citrus industry can survive until transgenic trees are commercially available. In part, this is due to Florida's citrus industry dependence on a minimal quantity of fruit production. Primarily, the industry produces orange juice in a few large processing plants. If HLB causes production to fall below a certain level, processing plants will have to close, with each closing resulting in a permanent reduction of the industry. Because no alternative crops are known that could replace citrus, HLB has the potential to eliminate the livelihood of a substantial proportion of people in Florida.

Thus, something needs to happen quickly to retain citrus production. As noted, the CTV vector is unusually stable. One approach that recently has become evident is the need to use the CTV vector in the field expressing anti-Las genes as an interim measure until transgenic trees are available. There are several advantages in using the CTV vector in the field compared with traditional transformation. The first is time. Transformation of citrus is produced through tissue culture and results in juvenile characteristics such as thorniness and lack of flowering and fruiting. It requires several years for the juvenile characteristics to disappear. Additionally, since transgenic plants go through tissue culture, somaclonal variation can occur in new plants. Before production of thousands of acres of a particular clone, field evaluation for fruit and juice quality and horticultural characteristics needs to be done. This takes years. The advantage of the CTV vector is that it can be graft-transmitted to the citrus varieties that have been grown for years, with known performance, with no change in phenotypes, thus avoiding both juvenile characteristics and the need for field evaluation. This could save a number of years.

Another advantage of the vector is that it could be applied to trees already in the field. The vector can be easily applied to field trees by graft inoculation. Trees that are not already infected with HLB could be protected. If genes are found that are sufficiently effective against Las, the biggest advantage of the CTV vector is that it could be used to treat trees already infected with HLB. The vector plus the anti-HLB gene could be used therapeutically to reverse the epidemic.

From a standpoint of effects on the environment, the vector puts nothing into the environment permanently. Whatever is inserted into CTV will be eventually lost. Based on greenhouse studies, the best guess is that somewhere between 70 and 80% of trees would retain the foreign sequence for perhaps 10 years. Afterwards, however, the inserted sequences will be recombined out and the plants would be infected with a CTV isolate identical to that which is already endemic in the state.

Thus, the potential uses of the CTV vector have changed with the circumstances. Initially it was considered only a laboratory tool. Now we are proposing to do things not even thought of three to four years ago, which poses new questions: what could a similar vector do for nut trees or stone fruit trees or even forest trees? Instead of battling diseases, could value-added attributes be added to trees? From an environmental safety standpoint the virus used as the basis of the vector already should be widely prevalent in the environment so that nothing new is permanently added to the ecosystem.

7 Finale

The major lesson learned so far from virus-based vector design is that building effective vectors requires understanding of the regulation of viral gene expression. It is fairly easy to insert a reporter gene into a favorite virus and monitor expression in protoplasts or for a limited time in portions of an herbaceous plant. It is much more difficult to create a vector that both expresses the inserted gene at a sufficient level and is stable for a long enough period to be useful. Clearly this technology is still in its infancy; future vectors will use the properties of a range of viruses and be facilitated by the expression of functional elements from the host genome as well as those of the virus. The ultimate vectors are not likely to be recognizable as viruses. In fact, a first step in this direction was the deconstruction of vectors into easily manipulatable subunits that are inoculated via agroinoculation, where the virus inoculum entered the cytoplasm from the nucleus and the final stages of vector construction were done by in vivo recombination (to be described in successive chapters of this book). However, the vector could be refined much further. An example is a question I burdened my students with when I taught Molecular Plant Virology almost twenty years ago:

> "Your job is to build a vector that (1) must replicate and move systemically throughout a plant; (2) must insert its target genes into the plant genome; (3) and must completely disappear, leaving the target sequence in the plant genome. Your toolbox contains all viruses and all known biological products. Remember, a vector is not a virus; it is a device designed to perform a specific function."

I am not aware of any work attempting to produce a vector that would integrate genes into the plant genome as described. In medicine, this would be called 'gene therapy.'

Virus-based vectors have been available now for a quarter of a century—forever in terms of molecular biology time. The web site, www.plantpharma.org/resources/patents/plant-viral-systems/, lists more than 60 plant viral vector patents. They have become standard technology in research laboratories. However, their proposed commercial potential has not yet been reached. Because of their capability of producing large amounts of commercial specialty products, which could utilize large tracts of land, it was expected that plant virus-based vectors would open up new applications for agriculture. So far, this expectation has not been met. The major reason is that the initial expectations were somewhat naive. Scientists tend to have big dreams, but rarely do they have experience in processing or marketing. However, the biggest hurdle for commercial development of agriculture based on virus-based vectors is choosing the appropriate product. There are a limited number of products that can efficiently be made in plants using vectors. There are a limited number of value-added products that are commercially viable. Only products that fall within both groups can be profitable. Most of the products that have been considered are pharmaceuticals or diagnostic. Maybe attention should be focused in other directions. Why not industrial enzymes that do not need to be purified—perhaps to make biofuels? Why not make better fruit or nuts—with more vitamins or more healthful chemicals?

In the final analysis, we come back to the definitions of basic and applied research—the vector is the solution, what problems can it solve?

Acknowledgments I thank Dr. Ben Graffam for editorial suggestions and the initial quotation.

References

Agranovsky AA (1996) Principles of molecular organization, expression, and evolution of closteroviruses: over the barriers. Adv Virus Res 47:119–158

Albiach-Martí MR, Mawassi M, Gowda S, Satyanarayana T, Hilf ME, Shanker S, Almira EC, Vives MC, López C, Guerri J, Flores R, Moreno P, Garnsey SM, Dawson WO (2000) Sequences of citrus tristeza virus separated in time and space are essentially identical. J Virol 74:6856–6865

Ayllon MA, Gowda S, Satyanarayana T, Karasev AV, Adkins S, Mawassi M, Guerri J, Moreno P, Dawson WO (2003) Effects of modification of the transcription initiation site context on citrus tristeza virus subgenomic RNA synthesis. J Virol 77:9232–9243

Ayllón MA, Gowda S, Satyanarayana T, Dawson WO (2004) *Cis*-acting elements at opposite ends of the *Citrus tristeza virus* genome differ in initiation and termination of subgenomic RNAs. virology 322:41–50

Ayllón MA, Satyanarayana T, Gowda S, Dawson WO (2005) An atypical 3′-controller element mediates low-level transcription of the p6 subgenomic mRNA of *Citrus tristeza virus*. Mol Plant Pathol 6:165–176

Bar-Joseph M, Garnsey SM, Gonsalves D (1979) The closteroviruses: a distinct group of elongated plant viruses. Adv Virus Res 25:93–168

Beck DL, Dawson WO (1990) Deletion of repeated sequences from tobacco mosaic virus mutants with two coat protein genes. virology 177:462–469

Chapman S, Kavanagh T, Baulcombe D (1992) *Potato virus X* as a vector for gene expression in plants. Plant J 2:549–557

Culver JN, Lehto K, Close SM, Hilf ME, Dawson WO (1993) Genetic position affects the expression of tobacco mosaic virus movement and coat protein genes. Proc Natl Acad Sci USA 90:2055–2059

Dawson WO, Bubrick P, Grantham GL (1988) Modification of the tobacco mosaic virus coat protein gene affects replication, movement and symptomatology. Phytopathol 78:783–789

Dawson WO, Lehto K (1990) Regulation of tobamovirus gene expression. Adv Virus Res 38:307–342

Dawson WO, Lewandowski DJ, Hilf ME, Bubrick P, Raffo AJ, Shaw JJ, Grantham GL, Desjardins PR (1989) A tobacco mosaic virus-hybrid expresses and loses an added gene. virology 172:285–292

Dolja VV, Kreuze JF, Valkonen JP (2006) Comparative and functional genomics of closteroviruses. Virus Res 117:38–51

Donson J, Kearney CM, Hilf ME, Dawson WO (1991) Systemic expression of a bacterial gene by a tobacco mosaic virus-based vector. Proc Natl Acad Sci USA 88:7204–7208

Folimonov AS, Folimonova SY, Bar-Joseph M, Dawson WO (2007) A stable RNA virus-based vector for citrus trees. virology 368:205–216

Folimonova SY, Alexey S, Folimonov AS, Tatineni S, Dawson WO (2008) *Citrus tristeza virus*: survival at the edge of the movement continuum. J Virol 82:6546–6556

Folimonova SY, Robertson CJ, Shilts T, Folimonov AS, Hilf ME, Garnsey SM, Dawson WO (2010) Strains of *Citrus tristeza virus* does not exclude superinfection by other strains of the virus. J Virol 84:1314–1325

French R, Janda M, Ahlquist P (1986) Bacterial gene inserted in an engineered RNA virus: efficient expression in monocotyledonous plant cells. Sci 231:1294–1297

Gowda S, Satyanarayana T, Ayllón MA, Albiach-Martí MR, Mawassi M, Rabindran S, Garnsey SM, Dawson WO (2001) Characterization of the *cis*-acting elements controlling subgenomic mRNAs of *Citrus tristeza virus:* production of positive- and negative-stranded 3′-coterminal and positive-stranded 5′-coterminal RNAs. virology 286:134–151

Gowda S, Ayllón MA, Satyanarayana T, Bar-Joseph M, Dawson WO (2003) Transcription strategy in a *Closterovirus*: a novel 5′-proximal controller element of *Citrus tristeza virus* produces 5′- and 3′-terminal subgenomic rnas and differs from 3′-ORF controller elements. J Virol 286:134–151

Hilf ME, Dawson WO (1993) The tobamovirus capsid protein functions as a host-specific determinant of long-distance movement. virology 193:106–114

Hilf ME, Karasev AV, Pappu HR, Gumpf DJ, Niblett CL, Garnsey SM (1995) Characterization of *Citrus tristeza virus* subgenomic RNAs in infected tissue. virology 208:576–582

Karasev AV (2000) Genetic diversity and evolution of closteroviruses. Annu Rev Phytopathol 38:293–324

Karasev AV, Boyko VP, Gowda S, Nikolaeva OV, Hilf ME, Koonin EV, Niblett CL, Cline K, Gumpf DJ, Lee RF, Garnsey SM, Lewandowski DJ, Dawson WO (1995) Complete sequence of the *citrus tristeza virus* RNA genome. virology 208:511–520

Karasev AV, Hilf ME, Garnsey SM, Dawson WO (1997) Transcriptional strategy of closteroviruses: mapping the 5′ termini of the *citrus tristeza virus* subgenomic RNAs. J Virol 71:6233–6236

Kearney CM, Donson J, Jones GE, Dawson WO (1993) Low level of genetic drift in foreign sequences replicating in an RNA virus in plants. virology 192:11–17

Kumagai MH, Turpen TH, Weinzettl N, della-Cioppa G, Turpen AM, Donson J, Hilf ME, Grantham GL, Dawson WO, Chow TP, Piatak M Jr, Grill LK (1993) Rapid, high-level expression of biologically active alpha-trichosanthin in transfected plants by an RNA viral vector. Proc Natl Acad Sci USA 90:427–430

Kumagai MH, Donson J, della-Cioppa G, Harvey D, Hanley K, Grill LK (1995) Cytoplasmic inhibition of carotenoid biosynthesis with virus-derived RNA. Proc Natl Acad Sci USA 92:1679–1683

Lehto K, Dawson WO (1990a) Changing the start codon of the 30 K gene of tobacco mosaic virus from weak to strong does not increase expression. virology 174:169–176

Lehto K, Dawson WO (1990b) Replication, stability, and gene expression of tobacco mosaic virus mutants with a second 30 K ORF. virology 175:30–40

Lehto K, Grantham GL, Dawson WO (1990) Insertion of sequences containing the coat protein subgenomic RNA promoter and leader in front of the tobacco mosaic virus 30 K ORF delays its expression and causes defective cell-to-cell movement. virology 174:145–157

Lu R, Folimonov A, Shintaku M, Li W-X, Falk BW, Dawson WO, Ding S-W (2004) Three distinct suppressors of RNA silencing encoded by a 20-kb viral RNA genome. Proc. Natl. Acad. Sci. USA 101:15742–15747

Mawassi M, Miekiewska E, Hilf ME, Ashoulin L, Karasev AV, Gafny R, Lee RF, Garnsey SM, Dawson WO, Bar Joseph M (1995) Multiple species of defective RNAs in plants infected with citrus tristeza virus. virology 214:264–268

Pappu HR, Karasev AV, Anderson EJ, Pappu SS, Hilf ME, Febres VJ, Eckloff RMG, McCaffery M, Boyko V, Gowda S, Dolja VV, Koonin EV, Gumpf DJ, Cline KC, Garnsey SM, Dawson WO,

Lee RF, Niblett CL (1994) Nucleotide sequence and organization of eight 3′ open reading frames of the citrus tristeza closterovirus genome. virology 199:35–46
Pleij CWA, Abrahams JP, van Belkum A, Rietveld K, Bosch L (1987) The spatial folding of the 39 noncoding region of aminoacylatable plant viral RNAs. In: [a]Positive strand RNA viruses. In: Brinton M, Rueckert R (eds) 299–316. Alan R. Liss Inc., NewYork
Rabindran S, Dawson WO (2001) Assessment of recombinants that arise from the use of a TMV-based transient expression vector. virology 284:182–189
Satyanarayana T, Gowda S, Boyko VP, Albiach-Martí MR, Mawassi M, Navas-Castillo J, Karasev AV, Dolja V, Hilf ME, Lewandowski DJ, Moreno P, Bar-Joseph M, Garnsey SM, Dawson WO (1999) An engineered closterovirus RNA replicon and analysis of heterologous terminal sequences for replication. Proc Natl Acad Sci USA 96:7433–7438
Satyanarayana T, Gowda S, Mawassi M, Albiach-Martí MR, Ayllón MA, Robertson C, Garnsey SM, Dawson WO (2000) Closterovirus HSP70 homolog and p61 in addition to both coat proteins function in efficient virion assembly. virology 278:253–265
Satyanarayana T, Bar-Joseph M, Mawassi M, Albiach-Martí MR, Ayllòn MA, Gowda S, Hilf ME, Moreno P, Garnsey SM, Dawson WO (2001) Amplification of *Citrus tristeza virus* from a cDNA clone and infection of citrus trees. virology 280:87–96
Scholthof HB (1999) Rapid delivery of foreign genes into plants by direct rub-inoculation with intact plasmid DNA of a tomato bushy stunt virus gene vector. J Virol 73:7823–7829
Siegel A (1983) RNA viruses as cloning vehicles. Phytopathol 73:775
Shivprasad S, Pogue GP, Lewandowski DJ, Hidalgo J, Donson J, Grill LK, Dawson WO (1999) Heterologous sequences greatly affect foreign gene expression in tobacco mosaic virus-based vectors. virology 255:312–323
Takamatsu N, Ishikawa M, Meshi T, Okada Y (1987) Expression of bacterial chloramphenicol acetyltransferease gene in tobacco plants mediated by TMV-RNA. EMBO J 6:307–311
Tatineni S, Robertson CJ, Garnsey SM, Bar-Joseph M, Gowda S, Dawson WO (2008) Three genes of citrus tristeza virus are dispensable for infection and movement throughout citrus trees. Virology 376:297–307
Van Vloten-Doting L, Bol JF, Cornelissen B (1985) Plant virus-based vectors for gene transfer will be of limited use because of the high error frequency during viral RNA synthesis. Plant Mol Biol 4:323–326

Virus-Derived ssDNA Vectors for the Expression of Foreign Proteins in Plants

Edward P. Rybicki and Darrin P. Martin

Abstract Plant viruses with ssRNA genomes provide a unique opportunity for generating expression vehicles for biopharming in plants, as constructs containing only the replication origin, with the replication-associated protein (Rep) gene provided in *cis* or in *trans*, can be replicationally amplified in vivo by several orders of magnitude, with significant accompanying increases in transcription and expression of gene(s) of interest. Appropriate replicating vectors or replicons may be derived from several different generic geminiviruses (family *Geminiviridae*) or nanoviruses (family *Nanoviridae*), for potential expression of a wide range of single or even multiple products in a wide range of plant families. The use of vacuum or other infiltration of whole plants by *Agrobacterium tumefaciens* suspensions has allowed the development of a set of expression vectors that rival the deconstructed RNA virus vectors in their yield and application, with some potential advantages over the latter that still need to be explored. Several modern applications of ssDNA plant vectors and their future potential will be discussed.

E. P. Rybicki · D. P. Martin
Institute of Infectious Disease and Molecular Medicine,
Cape Town, South Africa

E. P. Rybicki (✉)
Department of Molecular and Cell Biology,
University of Cape Town, PB Rondebosch, Cape Town,
Western Cape 7701, South Africa
e-mail: ed.rybicki@uct.ac.za

Current Topics in Microbiology and Immunology (2011) 375: 19–45
DOI: 10.1007/82_2011_185

Published Online: 29 October 2011

Contents

1 Introduction

1.1 ssDNA Plant Viruses

Single-stranded (ss-) DNA viruses are spread among all domains of life, infecting prokaryotes (*Microviridae*, *Inoviridae*), animals (*Anelloviridae*, *Circoviridae* and *Parvoviridae*), as well as plants (*Geminiviridae*, *Nanoviridae*). All of these viruses, as well as a number of plasmids of bacteria and phytoplasmas, share a common replication mechanism: this is rolling circle replication (RCR), mediated by a virus or plasmid-encoded replication-associated protein. They also probably share a common origin, which may well be ancient and bacterial (Koonin and Ilyina 1992), as essential elements of these protein sequences appear to have been conserved across all sequenced examples (Fig. 1). The increasing potential of the plant viruses as vectors may well extend to use of other members as well—and in this respect it is interesting that a variety of phytoplasma-derived extrachromosomal DNAs encode geminivirus-like Reps (Nishigawa et al. 2001; Rekab et al. 1999); that a geminivirus-like mycovirus has been very recently discovered (Yu et al. 2010), and that a variety of algal- and otherwise-derived circo- and geminivirus-like sequences have been found in metagenomic screens of oceanic water samples (Nawaz-ul-Rehman and Fauquet 2009; Rosario et al. 2009).

Except for the parvoviruses, which have linear genomes, the ssDNA plant viruses are like all other ssDNA viruses in having circular genomes. They belong to the two taxonomic families *Geminiviridae* and *Nanoviridae*. The following section discusses basics of the molecular biology of the viruses that are important to understand for any subsequent discussion of their use as vectors for high-level expression of proteins in plants.

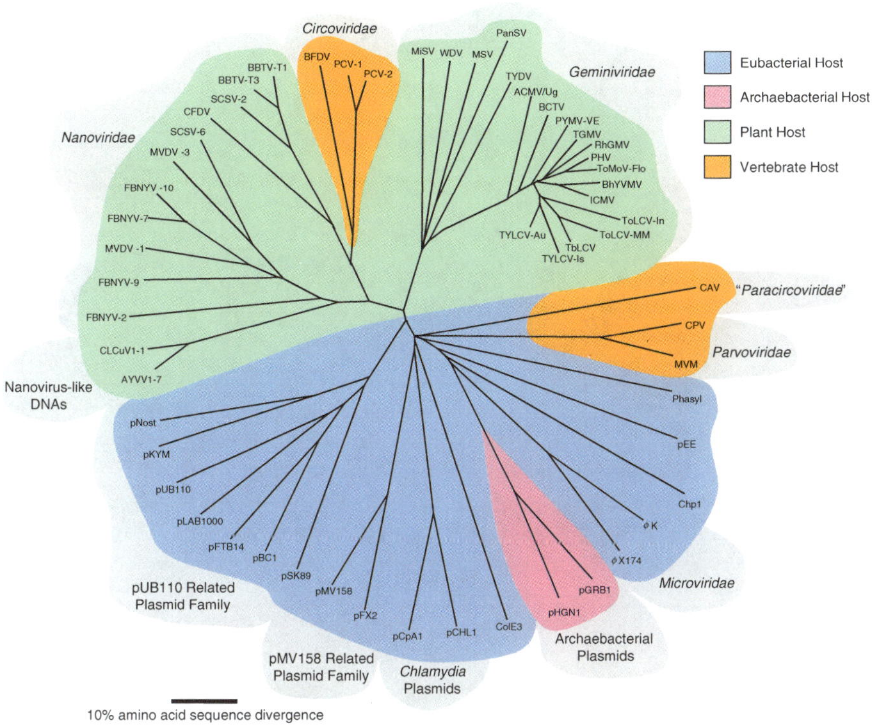

Fig. 1 The possible evolutionary relationships of rolling circle replication (RCR) domains within the replication-associated proteins encoded by a number of ssDNA replicons. Based on a published alignment of the $\sim$80 amino acids spanning the three RCR motifs from a diverse group of ssDNA replicons (Koonin and Ilyina 1992), this unrooted dendrogram was constructed by the neighbour joining method (Saitou and Nei 1987) using 1,000 bootstrap iterations. While bootstrap support for nodes near the centre of the dendrogram is relatively poor (all nodes with less than 30% support have been collapsed), support for all the presented nanovirus, circovirus and geminivirus groupings was in excess of 50%. There was 48% bootstrap support for the pUB110 plasmid family being more closely related to the geminivirus/circovirus/nanovirus grouping than to other prokaryotic replicons

1.1.1 Geminiviruses

Geminiviruses are divided into four genera; namely, *Mastrevirus*, *Begomovirus*, *Curtovirus* and *Topocuvirus*, based on genetic organisation and insect vectors. Mastre-, curto- and *topocuviruses* all have single-component genomes of 2.7–2.8 kb in size, while begomoviruses may have single- or two-component genomes, each of the same size (Stanley et al. 2005). Their genomes all have a characteristic organisation as shown (Fig. 2a), with ORFs diverging in virion (= V) and complementary (= C) senses from an intergenic region which contains elements of both V- and C- sense promoters, and the origin of replication (= *ori*). The *ori* in all cases contains a characteristic stem-loop forming sequence, with the

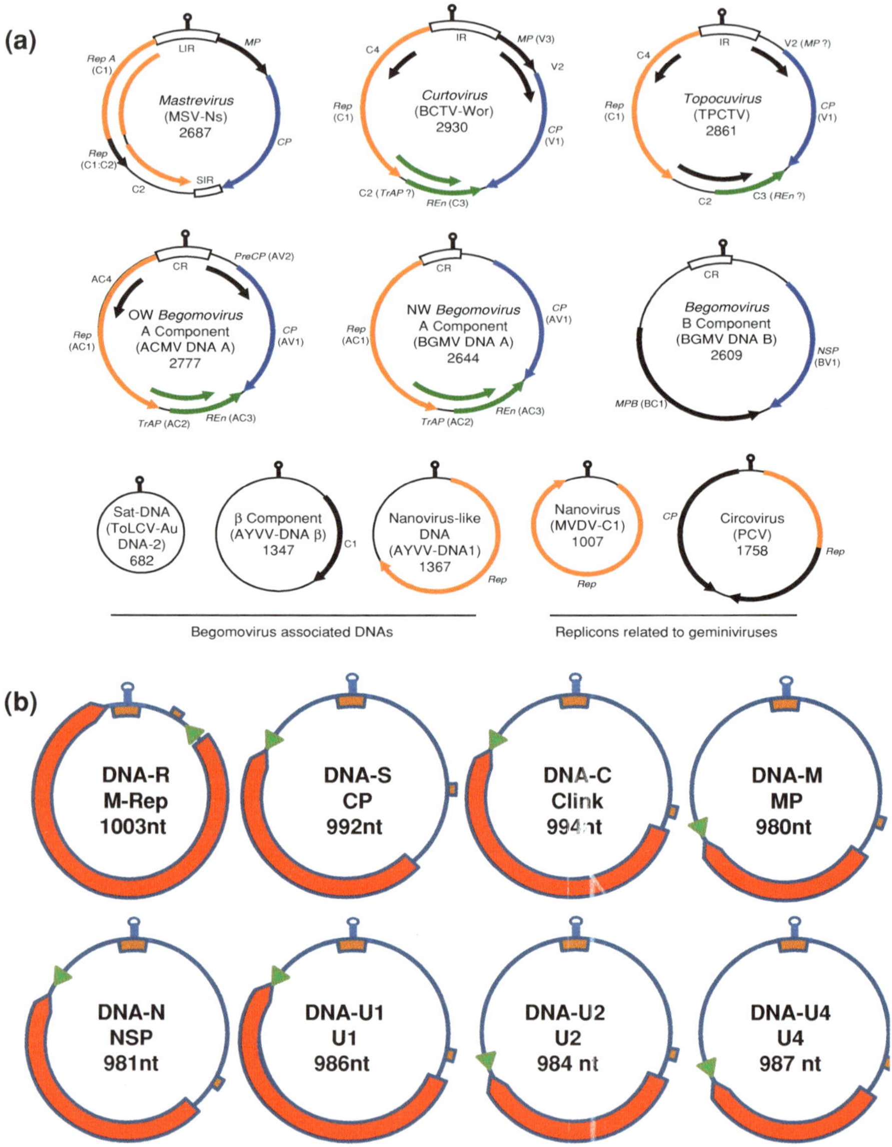

invariant nonanucleotide sequence TAATATTAC in the loop forming the genomic (+) strand origin of replication. A smaller intergenic region is found in mastreviruses, at the convergence of V and C-sense ORFs.

Other ssDNAs associated with certain Old World (OW) begomovirus infections (see Fig. 2a) are satellites (Dry et al. 1997; Mansoor et al.1999; Saunders and

◂**Fig. 2** **a** The genomic arrangements of geminiviruses and some other dependent replicons. Origins of (+) strand synthesis are indicated by the stem-loop symbol at 12 o'clock. OW Old World. NW New World. IR intergenic region. LIR long/large intergenic region. SIR short/small intergenic region. CR common region that is nearly identical in the A and B components of bipartite begomoviruses. Genes and parts of genes in the same colour either express or potentially express proteins with detectable sequence homology. *Rep* replication-associated/initiator protein gene found in all geminiviruses, circoviruses and nanoviruses. *RepA* variant of *Rep* potentially expressed from an unspliced complimentary sense transcript and encoding a multifunctional regulatory protein that is unique to the mastreviruses. *MP* movement protein gene of mastreviruses and curtoviruses. *CP* coat protein gene. *REn* replication enhancer gene found in begomoviruses and curtoviruses. *TrAP* transcription activator protein gene found in begomoviruses. *PreCP* pre-coat protein gene. *NSP* nuclear shuttle protein gene. *MPB* movement protein gene found on the B component of bipartite begomoviruses. **b** Genomic arrangement of a nanovirus: Faba bean necrotic stunt virus (FBNSV). Figure adapted from Grigoras et al. (2009). Genome organization of FBNSV. The eight proven genomic DNAs are designated according to current nomenclature (Vetten et al. 2005). The common region and internal stem-loop (CR-SL) sequence is shown in brown, with stem-loop above it. Protein-encoding ORFs are represented by *red arrows*. TATA boxes (*brown*) and polyadenylation signals (*green arrows*) are shown for each component. Sizes of the individual DNA components are shown within the components. M-Rep = Master Rep (33.2 kDa); CP = coat protein (19.2 kDa); Clink = cell cycle link protein (19.9 kDa); MP = movement protein (13.2 kDa); NSP = nuclear shuttle protein (17.6 kDa); U1- U4 = proteins of unknown function

Stanley 1999; Saunders et al. 2000), and one at least is derived from a nanovirus, but the β component DNA (Saunders et al. 2000) is a unique satellite DNA which can massively enhance the virulence of its associated begomovirus.

Geminiviruses are unique in having geminate or doubled particles, each consisting of two partially assembled T = 1 capsids joined at the missing pentameric vertex, and each encapsidating a single genome component (Bottcher et al. 2004; Zhang et al. 2001) (Fig. 3). There is a severe size constraint on the sizes of genomes that can be encapsidated, with ssDNAs of around 1.3–2.8 kb being strongly favoured, although multiples of the normal genome size can be incorporated into multimeric particles (Casado et al. 2004; Frischmuth et al. 2001).

The geminiviruses are obligately transferred between plants by insect vectors in natural infections: while some of the viruses have been mechanically transmitted, this is usually done only with great difficulty and recourse to methods which mimic injection via piercing mouthparts. In some cases it has been shown that cloned dsDNAs are infectious, but this is only true for dicot hosts. The begomoviruses are vectored by one species of whitefly (*Bemisia tabaci*), and may be the largest single group of characterised plant viruses; they have a world-wide distribution, while only infecting dicotyledonous plants. Interestingly, New World (NW) begomoviruses are very predominantly two-component, while Old World (OW) viruses may be one- or two-component and are often associated with satellite DNAs, usually β-components (see Briddon and Stanley 2006; Nawaz-ul-Rehman and Fauquet 2009). Mastreviruses are vectored by different species of cicadellid leafhoppers, and have an "Old World" distribution only: most characterised members infect grasses, and Maize streak virus (MSV) is important in maize in Africa and Wheat dwarf virus (WDV) in cereals in Eurasia. However, some mastreviruses have adapted to infect dicots, and Tobacco yellow dwarf (TYDV), Bean yellow dwarf (BeYDV) and Chickpea chlorotic dwarf viruses (ChCDV) are increasingly being implicated in crop diseases worldwide (Thomas et al. 2010). Curtoviruses are vectored by leafhoppers, and topocuviruses by planthoppers. Both curto and topocuviruses have a limited distribution, only a few identified members, and infect only dicots (Stanley et al. 2005).

While specific geminiviruses have relatively limited host ranges, experimental host range may be considerably widened: for example, while African cassava mosaic begomovirus (ACMV) does not naturally infect tobacco (*N. tabacum*) or most certainly not maize, the A genome component replicates autonomously in protoplasted cells of both plants after transfection of partially dimeric DNA constructs (Paszkowski et al. 1993). Of interest to its potential use as a vector in heterologous systems, however, was that insertion of 1,165 bp of foreign DNA downstream of the *cp* promoter abolished replication in maize cells, but not in tobacco.

1.1.2 Nanoviruses

There are two genera of nanoviruses; namely, *Nanovirus* and *Babuvirus*. The former includes a number of viruses infecting mainly legumes, such as Faba bean necrotic yellows virus (FBNYV), Milk vetch dwarf virus (MDV), and Subterranean clover stunt virus (SCSV). The latter genus consists of one well-described species of virus, namely Banana bunchy top virus (BBTV), and two more putative members in Abaca bunchy top virus (ABTV) and Cardamom bushy dwarf virus (CBDV) (www.ictvonline.org). Viruses in both genera are vectored by aphids.

The viruses have genomes consisting of multiple components of around 1 kb in size. These all have the same basic genetic organisation, with a single ORF in the virion (V) sense adjacent to a "common region stem-loop" or CR-SL sequence, very similar to geminiviruses, shared between all components of a given virus (Fig. 2b). Five genome components are common to nano- and babu-viruses: DNA-R encodes a master Rep (MRep) protein; DNA-S encodes the CP; DNA-C encodes Clink, a cell cycle affecting protein which is known to interact with retinoblastoma (Rb) protein analogues (Lageix et al. 2007); DNA-M encodes a movement protein, and DNA-N a nuclear shuttle protein (NSP). There are three other DNAs (DNAs-U1, -U2, and -U4), which encode as yet uncharacterised proteins, found in the nanoviruses FBNYV and MDV. Another DNA (DNA-U3) has been identified from the babuviruses BBTV and ABTV (Grigoras et al. 2009; Gronenborn, 2004).

It is presumed that nanovirus particles are complete $T = 1$ icosahedra, encapsidating only one genome component each (Harding et al. 1991; Thomas and Dietzgen 1991).

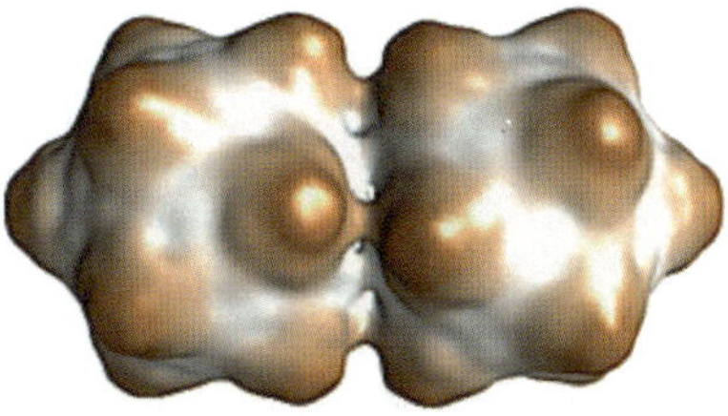

Fig. 3 Three-dimensional reconstruction of a geminate particle from cryoelectron microscopy data. Geminate particle of Maize streak virus: note characteristic doubled T = 1 particles, paired at missing vertex. [Adapted from Shepherd et al. (2010). Reconstruction by Kyle Dent and Trevor Sewell; copyright by the Electron Microscope Unit, University of Cape Town]

1.2 Replication of the Viruses

An understanding of the process of replication of the viruses aids in understanding vector construction, and possibly the uses to which elements of the virus genomes may be put. The following section discusses in some detail aspects of the replication of different gemini- and nanoviruses relevant to the purposes of this review.

The virions are injected via their vector insect's piercing mouthparts directly into cells of the host plant, where they presumably wholly or partially uncoat, and the single-stranded genomes are localised via cell machinery into the nuclei in a process that may be mediated by associated coat protein (CP) molecules. The ssDNA is converted to episomal dsDNA by host polymerases, most probably those involved in repair functions. Early transcription produces mainly mRNA for replication-associated protein (Rep), which serves to nick the double-stranded replication intermediates at the *ori* so as to allow covalent binding of Rep to the free 5′ residue. This allows single-strand displacement (rolling circle replication, or RCR) by a processive DNA polymerase adding onto the free 3′ end of genomic DNA. The Rep then acts to recircularise the displaced linear ssDNA genomes by ligation, aided by their circularisation due to annealing of the stem-loop region. Encapsidation of genomes is probably due to interaction of transiently sssDNA with accumulating CP, which self-localises to the nucleus via nuclear localisation signals, and subsequent sequestration by assembly (reviewed by Gronenborn 2004; Gutierrez 1999, 2000; Hanley-Bowdoin et al. 2000; Palmer and Rybicki 1998).

1.2.1 Geminivirus Replication

For all geminiviruses, the 5′-TAATATT↓AC-3′ sequence in the mastre-, topocu-, curto- and single component begomovirus large intergenic region (LIR) and two-component begomoviral common region (CR) genomic strand of a dsDNA replicative intermediate is the site of action of Rep: this nicks at the position shown, and covalently binds the 5′-A to allow strand displacement by processive (presumably repair) host polymerase, and subsequent RCR. In all geminiviruses

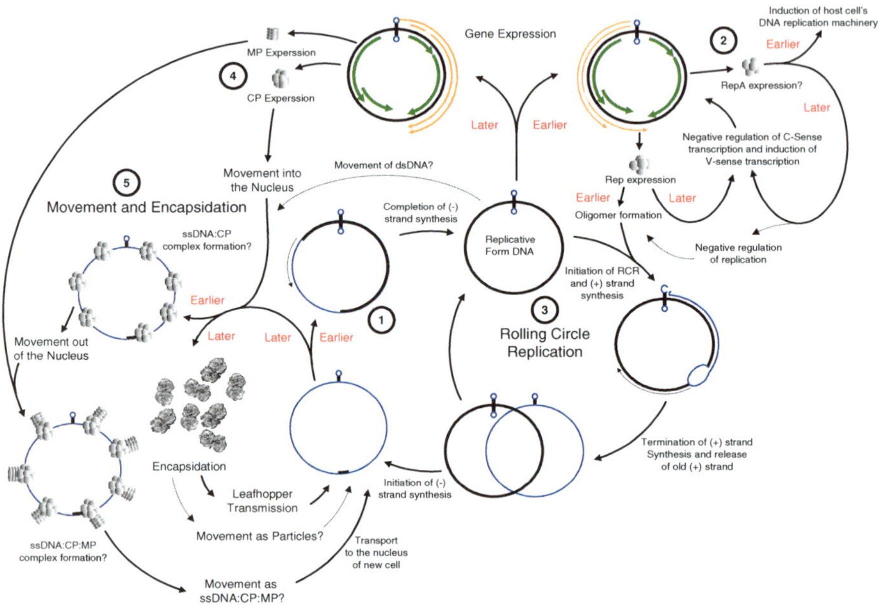

Fig. 4 Summary of the geminivirus infection and replication process, with Maize streak mastrevirus as an example. Early during an infection following the synthesis of a dsDNA replicative form; RF (**a**) RepA is most likely expressed and induces a cellular state in which viral DNA replication can occur (**b**) Rep is also expressed early and rolling circle replication begins (**c**) At a later point in the infection process, following genome amplification and possibly Rep and/or RepA induction of the V-sense promoter, MP and CP are expressed (**d**) and movement and encapsidation occur (**e**) Represented here is movement of unencapsidated ssDNA but it should be noted that it is possible that dsDNA and/or encapsidated ssDNA may also be moved either cell to cell or systemically within the phloem of plants. Whereas the involvement of MSV CP and MP in movement has been demonstrated (Kotlizky et al. 2000; Liu et al. 1997, 1999), the mechanics of the process are obscure and certain details have been borrowed from a cell to cell movement model proposed for the begomovirus, SqLCV (Lazarowitz and Beachy, 1999; Qin et al. 1998). While the probable timing of events is indicated, it is unlikely, for example, that absolutely no MP and CP expression occurs during the earlier stages of the infection process. ssDNA is represented by *blue lines*, dsDNA by *bold black lines* and RNA by *orange lines*

except the mastreviruses, Rep is expressed from a single (C1) ORF. In mastreviruses, Rep is expressed from the post-translationally spliced transcript of ORFs C1 and C2; RepA is a variant of Rep expressed from the C1 ORF only in an unspliced complementary sense transcript, and encodes a multifunctional regulatory protein that is unique to the mastreviruses. These are the only replication-associated proteins in mastreviruses. However, there is a REn or replication enhancer gene found in begomoviruses and curtoviruses which also has a homologue in topocuviruses, but its function in these viruses is not proven. Similarly, the well-characterised TrAP or transcription activator protein gene found in begomoviruses also has a possible homologue in curtoviruses (Hormuzdi and Bisaro 1995). The PreCP or pre-coat protein gene in OW begomoviruses is apparently involved in ssDNA accumulation

(Wartig et al. 1997), and in curtoviruses the V2 gene is apparently involved in regulation of ssDNA accumulation (Hormuzdi and Bisaro 1993; Stanley et al. 1992).

The actual process of genome replication is illustrated for mastreviruses in Fig. 4 given that these viruses have recently received the most attention as pharming vectors and will be featured in this review. However, the only DNA sequences required for a vector are *rep* and the LIR/CR for begomo, curto and topocuviruses, and *rep* plus the LIR and short intergenic region (SIR) for mastreviruses, if the *rep* expression is driven by a heterologous promoter (Palmer and Rybicki 1997). Host specificity is also less important than for viruses, as expression and replication may occur in a wide range of cell types without a movement constraint: for example, Bean yellow dwarf mastrevirus (BeYDV) with Maize streak virus (MSV) V region genes could replicate in and produce virus particles in *Nicotiana tabacum* protoplasts, but could not infect plants (Liu et al. 1999); the ACMV A genome component replicated successfully in protoplasted maize cells (Paszkowski et al. 1993).

Indeed, a variety of geminivirus-derived replicons also replicate in *A. tumefaciens*, purely as a result of Rep being expressed from cryptic bacterial promoters in the geminiviral LIR in partially redundant clones (Selth et al. 2002). There is also evidence that Ageratum yellow vein begomovirus (AYVV) replicons can be released from monomeric clones in a M13-derived vector in *E. coli* (Wu et al. 2007), and that Indian mung bean yellow mosaic begomovirus (IMBYMV) DNA-A replicates autonomously in budding yeast cells (Raghavan et al. 2004).

1.2.2 Nanovirus Replication

The overall process is much the same as with geminiviruses, with all nanoviruses having a "master Rep" which initiates RCR in the CR-SL region in the conserved loop sequence TA(T/G)TATT↓AC (T in babuviruses; G in nanoviruses). However, whereas with geminiviruses the Rep or RepA proteins interact with cell cycle regulatory proteins as well (see Gronenborn 2004), in familial nanoviruses this function is fulfilled by Clink, which has been shown to bind an Rb analogue as well as SKP1, part of the ubiquitin-protein turnover pathway (Aronson et al. 2000; Lageix et al. 2007).

2 ssDNA Viruses as Vectors: Historical

2.1 Geminiviruses

The use of viruses for the expression of foreign proteins in plants in fact predates the concept of using plants to express pharmaceutically-relevant proteins: while the first "pharming" was done by Hiatt et al. (1989) with the successful expression of a whole monoclonal antibody (mAb) in transgenic tobacco, the first use of a plant viral vector was by Takamatsu et al. (1987), who expressed a bacterial chloramphenicol acetyltransferase (CAT) gene in tobacco (*Nicotiana tabacum*) plants via inoculation of in vitro transcribed recombinant Tobacco mosaic virus (TMV) genomes. There was at the time, however, a significant problem in using the cloned genomes of plant DNA viruses to infect plants, which retarded their use as gene vectors: this was the general lack of infectivity of the DNA. This was surmounted in an ingenious use of the ability of *A. tumefaciens* to transfer DNA into plant cells by Grimsley et al. (1986), who demonstrated that complete head-to-tail dimeric clones of the dsDNA Cauliflower mosaic virus (CaMV) genomic DNA in *A. tumefaciens* were infectious if inoculated onto turnip stems, a process that became known as "agroinoculation" or sometimes "agroinfection". The technique relied on the fact that complete or even partially dimeric tandem viral genomic inserts in host cells could recombine to form circular genomes—which were infectious. This was soon applied to geminiviruses, with the demonstration that despite the fact that maize is not a host for *A. tumefaciens*, tandem copies of Maize streak virus (MSV) DNA were infectious when the bacterium was injected into maize seedlings—the first demonstration of the infectivity of cloned MSV (Grimsley et al. 1987). The technology subsequently spread swiftly through the geminivirus research community, with a number of demonstrations of infectivity for various generic geminiviruses (Boulton 1995; Briddon et al. 1989; Donson et al. 1988; Rochester et al. 1990).

Much of this early work was, despite its promise for high-value protein expression, mainly aimed at using geminiviruses as tools for probing viral replication and transport mechanisms, and host functions, and has been the subject of a number of comprehensive reviews (Stanley 1993; Timmermans et al. 1994).

The first use of geminiviruses to vector other genes in plants was by Ward et al. (1988), using African cassava mosaic virus (ACMV) with a coat protein (CP) gene replacement to vector CAT expression in *Nicotiana benthamiana*, and Hayes et al. (1988), who showed that Tomato golden mosaic virus (TGMV) DNA-A with the CP gene replaced could be used to express a bacterial enzyme—neomycin phosphotransferase (NPT)—either via release of a replicating DNA (replicon) from a transgenic tobacco with a partially dimeric DNA-derived insert, or via agroinoculation of transgenic tobacco with a partially dimeric DNA-B insert. In the latter case, the B genome would be mobilised by the self-replicating A genome, and in turn allow systemic spread of the latter by expression of movement proteins. However, the first properly systematic investigation of a recombinant geminivirus for expression of a foreign gene was probably by Hayes et al. (1989). This follow-up study showed that both NPT and β-glucuronidase (GUS) could be stably expressed at elevated levels relative to non-replicating controls, in plants transgenic for partially dimeric coat protein gene replacement TGMV DNA-A genomes, with or without the CP promoter being replaced by the CaMV 35S promoter, even if the DNA-A was significantly increased in size (Hayes et al. 1989). The constructs could also be used as "agroinfectious" vectors—introduced via recombinant *A. tumefaciens* inoculation—in the presence of DNA-B, but with a significant chance of vector deletion back to the wild-type size. The amplification of gene expression relative to the same constructs in single-gene copy transgenics under the 35S promoter was about 100-fold for both NPTII and GUS.

The potential use of mastreviruses for expressing foreign proteins was demonstrated by Kammann et al. (1991) who first established that it was possible to make a "shuttle vector" containing the *E. coli* ColE1 origin of replication and parts of the Wheat dwarf virus (WDV) genome, that could replicate in both bacteria and wheat cells. This was expanded by Matzeit et al. (1991), who demonstrated that replacement of the *cp* gene in the monomeric WDV shuttle vector by bacterial neomycin phosphotransferase, chloramphenicol acetyltransferase, and *β*-galactosidase genes did not affect the ability of the genome to replicate in protoplasts derived from *Triticum monococcum* cells, even though the *β*-galactosidase gene doubled the size of the vector.

A very interesting use for mastreviruses was pioneered by Laufs et al. (1990), who used Wheat dwarf virus (WDV) to vector the maize transposon Activator (Ac) and derivatives into wheat, maize and rice protoplasts. Rapid and efficient excision of Ac was detected only if the virus could replicate; excision of non-autonomous Ds elements was also possible if transposase activity was provided in *trans*. This was followed by evidence that agroinfection of a MSV-derived replicon could be used to vector the Ds element into whole maize plants, with proof of transposition being restoration of a wild-type MSV replicon and subsequent symptom expression (Shen and Hohn 1992). Sugimoto et al. (1994) then showed that Miscanthus streak mastrevirus (MiSV) could be used to vector Ds carrying a payload of a hygromycin phosphotransferase gene into rice protoplasts, which resulted in stable nuclear transformation. Proof that replication of the vector was required for efficient excision and transposition was given by Wirtz et al. (1997), who used WDV to vector Ds into maize protoplasts. The first authors raised the point at the time of the potential of combining a transposon with a geminiviral replicon for plant molecular genetic engineering; however, this potential appears to have been superseded as other techniques such as *A. tumefaciens* transformation improved.

An important milestone for the use of mastreviruses was the demonstration by Shen and Hohn (1994) that MSV-based replicons could be used via agroinfection to vector expression of *β*-glucuronidase (GUS) in maize plants via agroinfection. Expression of GUS was amplified by up to tenfold compared to use of a non-replicating vector; however, the vectors replicated only in cells of leaf primordia which had been exposed to recombinant *A. tumefaciens* during the infection phase, and did not spread systemically.

Two investigations which stand out from the late 1990s were those of Palmer et al. (1999) using MSV, and by others using the mastrevirus Tobacco yellow dwarf virus; TYDV (Needham et al. 1998). In the first, MSV-derived replicons were generated in suspension-cultured Black Mexican sweetcorn cells after biolistic co-bombardment with partially dimeric DNA constructs encoding a variety of constructs expressing the bialaphos resistance gene (*bar*) for initial selection purposes. The effect on replication of increasing the genomic size was tested by including the CaMV 35S promoter plus *bar* only; CaMV 35S Pr and the 550-bp maize *adh* I intron and 68-bp TMV Ω RNA leader sequences upstream of *bar*; and a fusion between *bar* and *E. coli* glutathione reductase (*gor*) genes. The three recombinant viral vectors ranging in size from 2.7 to 4.8 kb replicated efficiently; replicons were structurally stable, replicated to copy numbers of over 500 per haploid genome, and were detected for up to 3 years after introduction, and 2 years without selection. Expression levels of recombinant proteins were between 4 and 80 times the best level achieved by conventional transformation, depending on the protein. The TYDV work looked at expression of GUS from the *uidA* gene incorporated into an integrated partially dimeric TYDV-derived construct which produced replicons in regenerated plants: this was interesting at the time because of the success in incorporating a complete and normally-expressing *rep* gene in a transgenic plant; it is significant because of later developments involving inducible TYDV-based replicons in transgenic plants.

The foundation for the use of geminiviral replicons for molecular pharming was probably cemented in the early 2000s, primarily with the publication of several papers describing the use of Bean yellow dwarf mastrevirus (BeYDV). In the first, Mor et al. (2003) describe a two-component system tested in biolistically transfected suspension- cultured tobacco NT-1 cells, with Rep supplied in *trans* with a dependent replicon containing the *cis*-acting sequences LIR and SIR (see Fig. 2a), and expressing jellyfish green fluorescent protein (GFP) or GUS genes under control of a CaMV 35S Pr. They noted significant enhancement of indicator gene expression—up to 40-fold—and established that the two-component system worked well, as well as showing that only the mastreviral Rep, and not RepA as well, was necessary for effective expression. In the second, and purportedly the first vaccine-related application, Hefferon and Fan (2004) essentially reiterated the first group's results, and also reported use of a BeYDV-derived replicon for plant codon use optimised *Staphylococcus* enterotoxin B protein (SEB) gene expression as a vaccine model: they reported up to 20-fold increases in SEB concentration for NT-1 cells containing replicating vector compared to cells bombarded with the *rep-* version only. However, the vaccine connection was unconvincing, as yields were not given, nor was the final vector construct described, and no immunological evidence was offered. In their second paper, Hefferon et al. (2004) used a BeYDV replicon in NT-1 tobacco and Ad-2 *Arabidopsis thaliana* suspension-cultured cells to show enhanced production ($\sim$20x) of, and apparently correct glycosylation of, a MAP-K protein from *Arabidopsis* and Potato virus X CP, respectively. They also purified sufficient MAP-K from cells for biochemical analyses. In what is possibly the only use of a monopartite begomovirus for the purpose of foreign gene expression, Tamilselvi et al. (2004) described an Ageratum yellow vein virus (AYVV)-derived "*E. coli*—plant shuttle vector", which was used to electrophoretically transfect *N. benthamiana* mesophyll-derived protoplasts, and biolistically transfect tobacco BY2 suspension-cultured cells. The authors describe the plasmid vector—which contained a monomeric *cp*$^-$ AYVV genome, and a pUC19 backbone—as allowing extrachromosomal replication in plant cells as well as in *E. coli,* under ampicillin selection.

While the investigations described above laid a useful foundation for use of gemini- and other ssDNA viruses as gene vectors for pharmaceutical production, there were important limitations to their application. First, while whole-plant expression of proteins was occasionally possible via replicating geminivirus genomes in transgenic plants, it proved very difficult to successfully regenerate plants with integrated replication-competent geminiviral genomes. This is not surprising in retrospect, given the potential for Rep and other viral proteins for interacting with cell cycle regulatory proteins (see Gutierrez et al. 2004; Hanley-Bowdoin et al. 2000, 2004). However, it meant that either only very rare integration events with few viral genomes could be used to express foreign proteins, or that transgenic cell or callus cultures had to be used (see Palmer et al. 1999). Second, delivery of the viruses into plants for transient expression either via agroinfection or as cloned dsDNA was necessarily only into a limited number of sites per plant: this meant that the viral genomes had to be able to spread out of inoculated tissue in order to be effective as vectors, with all of the limitations in payload this necessitated, given size limitations for cell-to-cell transport and the necessity of incorporating movement protein gene(s). Another problem that came to light when *trans*-complementing viral genomes—each encoding a function necessary for the other component—were investigated as infectious vectors, was that recombination back to wild type was very efficient and very quick (Palmer and Rybicki 2001). In such a case, autonomously-replicating and spreading wild-type genomes would quickly come to dominate the infection, reducing transgene expression to zero. Again, this is obvious in retrospect, as it is known that several different begomoviral B-genomes can be associated with a single A-genome in natural infections, after apparently obtaining the requisite CR sequence by recombination so as to allow strain-specific recognition by the A genome-encoded Rep (Briddon et al. 2010).

The potential of using geminiviruses for high-level transgene expression therefore mostly languished, while the use of RNA viruses increased (see Rybicki, 2009a; Yusibov et al. 2006). However, possibly the most successful application of geminivirus-mediated transgene expression technology was reported by Zhang and Mason (2006), who used an alcohol-inducible two-component (*rep* + replicon) system in doubly-transformed tobacco NT-1 cells and in transgenic potato plants, to express GFP and the Norwalk norovirus (NV) CP. Rep expression was driven by the *alc*A Pr, under control of a CaMV 35S Pr-driven *alc*R or alcohol response protein gene, while replicons were flanked by LIRs, with gene expression driven by a 2xCaMV 35S Pr with a Tobacco etch potyvirus (TEV) 5′ UTR translational enhancer. In doubly-transformed NT-1 cells induced with 0.1% ethanol, there was significant amplification of GFP and NV CP expression up to 4 days post induction. In potato plants, expression peaked at 7 days and declined thereafter, possibly due to silencing. It was evident that protein expression was not proportional to mRNA expression, indicating that saturation of protein synthetic machinery could result from too high a level of transgene expression.

2.2 Nanoviruses

By contrast to the case for geminiviruses, the use of cloned nanovirus genomes was minimal for any purpose other than discovery of genome functions, given that the full genome of any nanovirus was only assembled by 2009 (Grigoras et al. 2009), for a new nanovirus called Faba bean necrotic stunt virus (FBNSV) (see also Fig. 2b). Early work that showed potential, however, was that of Dugdale et al. (1998), who investigated the potential of BBTV promoters for enhancement of gene expression in biolistically bombarded tobacco and banana cell cultures, and in transgenic tobacco and banana. Interestingly, expression was best in tobacco cells, with two promoters (BBTV DNA-2 and -6) being twofold better and as good as the CaMV 35S Pr. In banana cells, activities of Prs from DNAs 4 and 5—corresponding to movement protein (M) and Clink, respectively (Wanitchakorn et al. 2000)—were highest, and DNA 1 (Rep) the least, which was considered appropriate given the requirement for larger amounts of MP and Clink than for Rep.

Possibly the first use of replicons derived from a nanovirus was the use of parts of the Faba bean necrotic yellows virus (FBNYV) genome, for which the eight cloned genome components analogous to the related FBNSV had been shown to be infectious via biolistic but not via aphid transmission (Aronson et al. 2002; Timchenko et al. 2006). This work used agroinoculation of the complete M-Rep-encoding DNA-R and DNA-C segments, together with constructs based on the DNA-C encoding Clink, to express GFP and GUS in *N. benthamiana* leaf disks or *Medicago trunculata* leaves and leaf disks: while M-Rep served to create replicons from and amplify all FBNYV-derived DNAs, Clink co-expression was found to significantly amplify protein expression (Aronson et al. 2002). In the case of GUS, this was $\sim 6\times$ that driven by CaMV 35S promoter alone; for both GFP and GUS expression levels were correlated with viral DNA levels. This work established the potential of nanoviruses to be used as expression vectors, and also showed that the minimal replicon would need *mrep* in *cis* or in *trans*, releasing a replicon comprising the CR-SL of any cognate genome segment, together with either a heterologous promoter driving just the gene of interest, or *clink* as well with a DNA-C-derived replicon with the native *clink* promoter.

3 ssDNA Viruses as Vectors: Current Practice

The modern use of these viruses for expression of pharmaceutically-relevant proteins mostly dates from the application of another technology: this was the application of "agroinfiltration", or the use of recombinant *Agrobacterium tumefaciens* for transient gene transfer to normal plants, by means of syringe or vacuum infiltration of whole leaves or plants (Fischer et al. 1999; Schob et al. 1997). This process enables the somatic transfer of T-DNA to a high percentage of the cells in the infiltrated leaf or leaves, where it may express its payload whether it remains episomal, or becomes integrated [see Rybicki (2009a), for a review of its modern use in vaccine protein expression].

The combination of agroinfiltration with agroinfection effectively turned the majority of dicot-infecting geminiviruses or nanoviruses into potential expression vectors for dicots—not, however, for monocots, as these are generally recalcitrant to agroinfiltration due to leaf morphology and *Agrobacterium*-related host range properties (Heath et al. 1997; Reavy et al. 2007). The take-up by the geminivirus and plant molecular biology research community was swift—but the main application was and is in further demonstrating and investigating the phenomenon of post-transcriptional gene silencing, or what is now known as siRNA-mediated gene silencing (e.g.: Fofana et al. 2004; Muangsan and Robertson, 2004; Rodriguez-Negrete et al. 2009; Turnage et al. 2002).

One of the first applications of the new technology for geminivirus-based expression of a foreign as opposed to a vaccine-relevant gene was in fact done in suspension cell or hairy root cultures on *Nicotiana glutinosa*, using BeYDV. Collens et al. (2007) used the two-component complementing vectors previously described, delivered via auxotrophic *A tumefaciens*, to express GUS. While the work was proof that a transient expression system was compatible with a geminivirus vector, it was a curiosity rather than a milestone, given only modest (2×) increase in recombinant protein yield.

An interesting excursion in the field was by Kim et al. (2007), with a curtovirus-derived vector: they used Beet severe curly top virus (BSCTV)-derived replicons to express GFP in *N. benthamiana* leaf disks infiltrated with recombinant *A. tumefaciens*. While their best GFP expression was not impressive—they claim a maximum of 1.5% of total soluble protein (TSP)—they did achieve a 3× higher expression level with a 35S Pr in BCTV compared to 35S control, a 3.2× enhancement of that expression if they used a Cassava vein mosaic caulimovirus (CaVMV) Pr instead, and a further 2.4× enhancement if they co-infiltrated leaf disks with *A. tumefaciens* expressing the silencing suppressor protein p19 from Tomato bushy stunt tombusvirus (TBSV). They also showed that whole leaf infiltration resulted in higher levels of GFP than leaf disk infiltration, that it was far easier than leaf disk inoculations, and concluded that "...we have demonstrated the feasibility of a BCTV replicon-based vector system, which could potentially be used in a wide variety of plant cells to express proteins of commercial or medical interest, or proteins for biochemical analysis." However, apart from one further report of a Beet curly top curtovirus (BCTV)-derived virus-induced gene silencing or VIGS vector (Golenberg et al. 2009), no further developments have occurred with curtovirus vectors.

The first vaccine-relevant development of the modern age, then, was the investigation by Huang et al. (2009) of the potential of an agroinfiltration-mediated BeYDV-based transient expression replicon system to express large amounts of GFP, Hepatitis B virus (HBV) core antigen (HBc), and Norwalk norovirus (NV) capsid protein (NVCP). They used very similar vectors to the earlier work with transgenic tobacco cells—that is, *rep*$^{-}$ replicons with a 35S Pr co-transformed with Rep-expressing constructs—but added single-component replicons, with *rep*

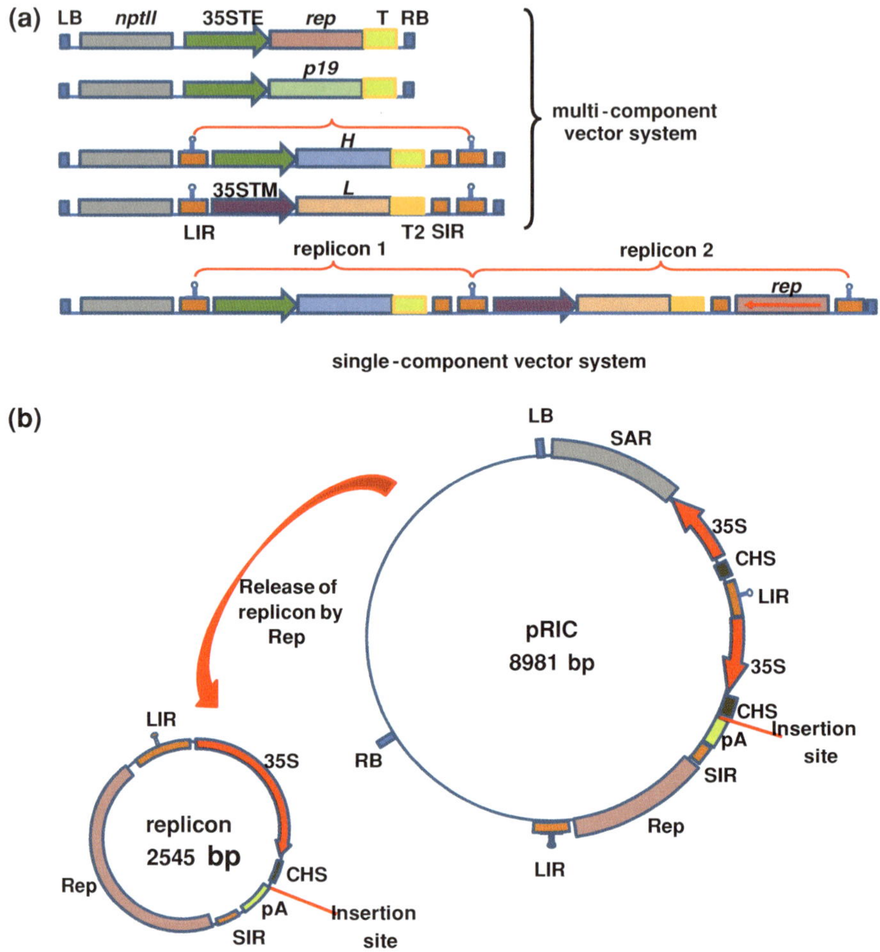

in its normal orientation between the right-hand LIR and the SIR at the 3′ end of the construct, as well as co-infiltration with p19-expressing vectors for silencing suppression. After rigorous optimisation, the authors succeeded with the single-vector system in producing HBc virus-like particles (VLPs) at 0.8 mg/g leaf fresh weight within 4 days of agroinfiltration—without p19. This was significant, as the previous highest plant yield of HBc was achieved using a two-component non-replicating system which incorporated p19 (Sainsbury and Lomonossoff 2008).

Further successes with the BeYDV system were achieved by the same group with full-size antibody expression (Huang et al. 2010), and by our group with expression of two human virus proteins (Regnard et al. 2010). In the case of antibody expression, it was possible to use either the three-component (*rep* + replicon + p19) system or a single-component, self-replicating vector system that released two replicons, to efficiently express both heavy and light

◀ **Fig. 5** Bean yellow dwarf mastrevirus (BeYDV)-derived replicon vectors. **a** Multiple and single vector multiple replicon constructs used for IgG expression and assembly. The schematic–adapted from the vectors created by Huang et al. (2010)—shows essential details of multigene and replicon systems suitable for the expression of several genes in one cell. The multicomponent system includes four separate plasmids, cointroduced via *A. tumefaciens* infiltration. All constructs encode a neomycin phosphotransferase II gene cassette for kanamycin resistance (*npt II*), and left and right border regions from A. tumefaciens T-DNA (LB + RB), transcription termination sequences (T + T2). 35STE CaMV 35S promoter with Tobacco etch potyvirus 5′ untranslated region (UTR) as translational enhancer; 35STM 35S Pr with Tobacco mosaic virus 5′ UTR. *rep* Rep gene (C1/C2 ORF from BeYDV). *p19* p19 silencing suppressor protein gene from. *H* IgG heavy chain gene. *L* IgG light chain gene. LIR long intergenic region of BeYDV genome, shown with stem-loop region; SIR short intergenic region. **b** Single vector and replicon constructs used for vaccine protein expression. The schematic shows the single-replicon, single-component vector described by Regnard et al. (2010) for expression of single genes. T-DNA is the sequence contained within left (LB) and right (RB) borders of the *A. tumefaciens* vector plasmid pTraC described by Maclean et al. (2007); in this system this also contains a CaMV 35S promoter driving a scaffold attachment region (SAR) protein. The replicon includes similar elements as described for Fig. 5a between two LIR regions. pA polyadenylation sequence; CHS chalcone synthase gene 5′-UTR. In addition, the system uses a Tomato spotted wilt tospovirus (TSWV) NSs silencing suppressor protein supplied in *trans* by a co-agroinfiltrated construct

chains of a full IgG molecule specific for Ebola virus, at levels up to 0.5 mg/g leaf tissue, or about 10× the standard transgenic plant yield of monoclonal antibodies (mAbs) (Rybicki 2009a) (see Fig. 5a). The authors rightly point out that this represents a significant advance in transient expression technology, as the otherwise state-of-the-art ICON Genetics RNA virus-derived vectors cannot be used similarly as single-gene-expressing replicon constructs compete, so that multiple component proteins cannot be expressed in the same cell. The authors speculate that "Our success in producing the fully assembled tetrameric functional IgG (two hetero-oligomeric subunits) with a two replicon single vector strongly suggests that simultaneous expression of as many as four hetero-subunits can be easily achieved using two of such vectors, or by creating single vectors with three or more tandem linked replicons".

The other recent success with BeYDV-based replicons was with another single-component self-replicating vector. This was a shuttle vector based on the mild strain of BeYDV sequenced by our laboratory (BeYDV-m; Halley-Stott et al. 2007). The vector comprises the native genome situated between two LIR copies, with the BeYDV *cp* and *mp* genes replaced with a 2xCaMV 35S promoter and transgene (Fig. 5b). The vector was tested by comparing it with a high-expressing non-replicating vector, pTraC, using transient cytoplasmic expression in *N. benthamiana* of the expression marker EGFP, and the genes encoding the subunit vaccine antigens, HPV-16 major CP L1 (HL1) and the HIV-1C p24 capsid protein (CA) antigen, all of which had previously been successfully expressed in plants using pTraC as well as other vectors (Maclean et al. 2007; Meyers et al. 2008). Expression vectors were co-infiltrated with *A. tumefaciens* expressing the NSs protein from Tomato spotted wilt tospovirus (TSWV), given that this markedly increases expression in our hands using these vectors. It was interesting at the outset that the HL1 vector replicationally

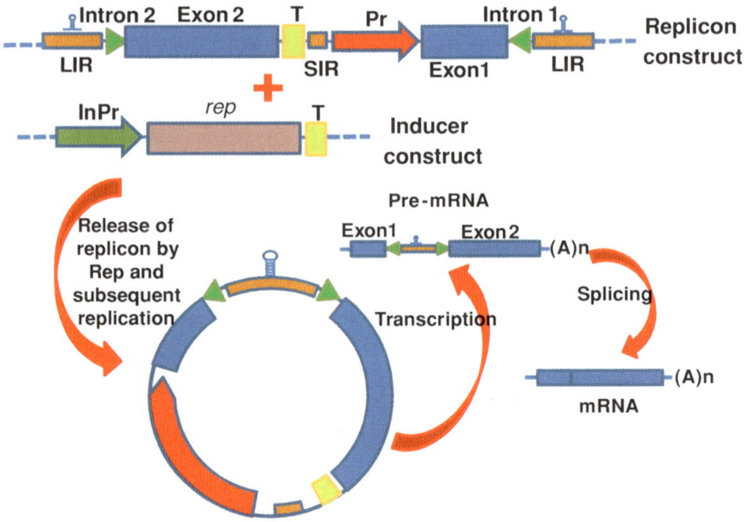

Fig. 6 InPact—inducible in-plant activation techology–transgenic vector schema. The diagram illustrates the broad concept of the InPact system: this relies on doubly transgenic plants with two independent insertions of the constructs shown. The inducer construct consists of the following components: InPr = inducible promoter (e.g.: alcohol-responsive transcriptional unit); *rep* = Rep gene of a ssDNA virus. The replicon construct has the payload gene divided into two exons (Exon 1/2), separated as shown, with a carefully-designed intron acceptor and intron donor sequence (Intron 1/2) flanking the LIR. *T* transcription terminator; SIR is shown for a mastrevirus-derived construct, but would not be required for other geminivirus-derived or nanovirus-derived constructs. Replicational release of the replicon by expression in trans of Rep, and subsequent circularisation, allows reconstitution of a payload gene with a LIR-containing intron. Transcription and splicing restores a mRNA with an undivided ORF as shown. Details taken from material obtained from JL Dale, Farmacule Bioindustries Pty Ltd, Victoria, Australia

released in *E. coli* and plants, but not in *A. tumefaciens*, given earlier experience with geminiviruses replicating in bacteria (see earlier, and Selth et al. 2002). Gene copy number in plants was assayed by quantitative PCR, and shown to increase for replicons by around 100x (HL1 gene) to 1000x (EGFP and p24 genes) by 3 days post-infiltration relative to non-replicating pTraC versions, and to remain steady for 7 days or more. Protein production peaked around 5 days for HL1, at around 1.5x that of pTraC; at 3 days for p24 at >3x, and was still increasing at 7 days for EGFP at >7x the pTraC level. These results are significant given that we have previously achieved high expression levels for EGFP and levels of HL1 of 0.5 g/kg wet weight using pTraC alone (Maclean et al. 2007), and 16 mg/kg for p24 targeted to the ER (Meyers et al. 2008).

While these developments are interesting and hold promise, possibly the best development in ssDNA virus-based expression technology has come from a company that is using it to make high-value proteins in plants via an inducible transgenic system. The company is Farmacule BioIndustries Pty Ltd of Victoria,

Australia (www.farmacule.com); the patented technology is called InPact, for "In-Plant Activation Technology", and relies on induced expression of Rep from one transgene to activate an independently inserted replicon construct. However, this replicon is constructed in such a way that, in the linear inserted form, the promoter is downstream from the gene of interest, which is itself divided so that expression from the gene is impossible unless the replicon is reconstituted into a circular form—and then only after splicing, as the whole LIR has been made into an intron (see Fig. 6). The technology allows tight control of expression of Rep—usually via alcohol induction—and near-absolute control over replicon release, and therefore of protein expression. Thus, transgenic plants potentially expressing even proteins that are highly toxic to them, or which would normally drastically affect development, can be grown normally until biomass is suitable for harvest, at which time the inducer is used to allow production of Rep and subsequently highly amplified expression of the replicon and protein (see Rybicki 2009b for a report). The system has been used successfully to produce high yields of vitronectin, a high-value protein used in tissue culture which is normally produced from bovine blood as well as enzymes like trypsin. The technology could potentially make use of any Rep-replicon pairing, but most development has been done using the indigenous mastrevirus TYDV. While development so far has apparently used the two-component system, it is of course possible to have plants transgenic for a replicon only, and introduce Rep via agroinfiltration if desired.

4 Conclusion and Future Prospects

The future use of geminivirus- or nanovirus-derived replicons for pharming is almost certainly very bright: the expression systems have been tested for a variety of geminiviruses in particular, and for a variety of plants, cells and types of proteins, and have generally been found to perform at least as well as conventional agroinfiltration, and considerably better in the recent examples shown above. Vector systems may be single component or be constituted as two or more components; while agroinfiltration may be used for whole plants, it is also possible to use it for leaves and even cell cultures; the use of transgenic plants are possible, especially if these are part of the InPact or similar system with expression only after induction, and so can transgenic cell lines be used over several years of culture. We can predict that the application of ssDNA vector and expression systems will take off exponentially in coming years, much as the use of deconstructed RNA virus vectors such as the MagnICON system of Icon Genetics has recently (Gleba et al. 2005, 2007).

The two main advantages of the technology for high-level pharmaceutical protein production over systems like the deconstructed RNA virus vectors are first, host range of the potential vectors, and second, the fact that it is not necessary to use two different viruses to express more than one polypeptide in the same cell, where heterodimeric or greater associations are need to produce a functional protein.

In the first place, the host range of ssDNA plant viruses is vast: if one only considers crop plants, the range includes such staples as maize, wheat, rice, banana, tomato, potato, cassava, sweet potato, sugar beets, cotton and beans. In one justification for the establishment of BCTV curtovirus as a silencing vector, Golenberg et al. (2009) point out that it infects 41 families of dicotyledonous plants in six subclasses—including several useful and well-characterised production hosts such as Fabaceae and Solanaceae. This is far greater than the essentially limited Tobacco mosaic virus strain normally used in "Magnifection", which is pretty much limited to *N. tabacum* and preferentially *N. benthamiana*—and while the developers are introducing other viruses with a wider range of hosts, such as Potato virus X (PVX) and potyviruses (see Rybicki 2009b), each of these is another challenge and another complicated design and sequence engineering job, whereas geminivirus vector development in particular so far has required little of this for good yields.

The second point was dealt with to some extent above in discussion of the work of Huang et al. (2010): they successfully managed to use co-agroinfiltration of two vectors to get individual leaf cells to produce both heavy and light chains of an IgG molecule, which was about as good as using a single vector in terms of yield. A TMV-based vector cannot produce more than one protein, and attempts to co-infiltrate more than one have failed due to competion between genomes—necessitating the use of non-interfering viruses (Giritch et al. 2006), with all of the immediate complexities that entails in terms of cloning strategies, and the added burden of vector design for maximal yield which will be beyond most laboratories.

Another useful feature of the DNA viruses is that they replicate and produce mRNA in the nucleus using cellular machinery, with all of the possible advantages that entails—such as the possibility of taking advantage of splicing, to limit expression of the transgene in the *A. tumefaciens*, if this is used; addition of polyA tails to limit mRNA degradation; the possibility of getting expression in transgenic plants at the time of choosing by means of a simple induction mechanism, and so on.

One possible drawback of using agroinfiltrated ssDNA replicons compared to MagnICON vectors is that presently, expression with the former is essentially limited to cells in which T-DNA transfer has occurred, while the latter maintain transport functions and can spread to untransfected or new cells. The advantage of this property is that, first, it serves to spread expression into effectively all leaf tissue in the plant; second, that it allows a significant reduction in the amount of *A. tumefaciens* that must be used for infiltration, given that the whole plant need not be infiltrated for maximal expression. This is not a trivial barrier to surmount, given that geminivirus and possibly nanovirus genome movement is complicated by there being a number of different proteins involved in the different genera of geminiviruses, let alone between the two families. For example, two-component NW begomoviruses exploit the B-encoded MP or BC1 ORF-encoded protein as a plasmodesmatal gate protein and the NSP or BV1 as a nuclear shuttle protein (Noueiry et al. 1994; Pascal et al. 1994). However, single-component begomo-, curto- and topocu- and mastreviruses have neither of these, and instead use A-type genome component encoded functions, such as the MP or V1 ORF and CP or V2 ORF proteins of mastreviruses, for movement between cells and nuclear transport respectively (Liu et al. 1997, 2001), or the C4 and/or V2 "pre-coat" ORF protein(s) for movement and CP for nuclear localisation in begomoviruses (Jupinet al. 1994; Padidam et al. 1996). Moreover, begomovirus genome movement between cells exhibits a strong size-dependency (Gilbertson et al. 2003; Rojas et al. 1998), and the inability of MSV-derived vectors with inserts to spread in plants has already been noted. Additionally, the thought of using transgenic plants expressing movement proteins to complement replicon spread is itself a problem, as the plants may have abnormal phenotypes and in fact be partially or substantially resistant to cognate virus infection and spread (Pascal et al. 1993; von Arnim and Stanley 1992).

Taken together these observations add up to a complex problem to be solved if one wishes ssDNA virus-derived replicon vectors to spread in agroinfiltrated or -inoculated plants. However, there are some obvious solutions if this is regarded as a problem, such as using transgenic plants incorporating the InPact technology, where expression is limited only by which tissue has been exposed to the inducing agent, or *mp*-transgenic plants which are not symptomatic, or in which MP expression is inducible, with agroinfiltration of replicons not exceeding genome size so that spread will occur.

It is also feasible to engineer ssDNA replicons in the same way that the deconstructed TMV-based vectors have been; that is, with codon optimisation, removal of cryptic splice sites, molecular breeding of various of their proteins to either remove unwanted properties, or enhance desired characteristics. Thus, it is possible that expression could be enhanced without adding movement functions, to the point where these are not considered necessary.

It is worth mentioning that while we have treated geminiviruses and nanoviruses as being essentially equivalent for purposes of this review, in terms of expression at least, they are probably not. For example, while considerable protein expression amplification was demonstrated using the mastreviruses MSV in cells, and BeYDV in cells and plants, the same was not true of the curtovirus BSCTV; nor have the begomoviruses that have been tested been particularly impressive in terms of yield of recombinant protein (see Palmer and Rybicki 1997). This may have to do with the fact that mastreviruses in both cereal plants and isolated cells appear to replicate to very high copy number compared to begomoviruses, and probably also curto- and topocuviruses: indeed, early workers found that WDV in protoplasts could replicate up to 30,000 genome copies per cell (Matzeit et al. 1991), a finding essentially replicated for MSV by Palmer et al. (1999), compared to values for begomoviruses of the order of 1,000 genomes/cell. This difference in cell and whole plant DNA content can be very simply demonstrated in protocols for extraction of replicative form dsDNA from infected plants: it is far easier to purify this from mastrevirus-infected maize, wheat or even beans, than it is to purify begomovirus-derived dsDNA. In fact, MSV dsDNA obtained from maize using a bacterial plasmid miniprep protocol is abundant and pure enough to be sequenced directly by the Sanger dideoxy method (Palmer et al. 1998). Thus, while it is possible to use any ssDNA plant virus as a vector, in practice the choice may be constrained by the simple fact of degree of genome amplification: the levels of mRNA and probably to a lesser extent of protein, given the capacity of geminiviral replicons to saturate the cell synthetic machinery, will inevitably be greater for a greater copy number.

In closing, it should be evident that there is ample room for further optimisation of ssDNA virus-based expression systems in plants: a few representatives of only two generic sets of geminiviruses have been tested out of four sets and hundreds of sequenced genomes available; additionally, familial nanoviruses are almost completely unexploited in terms of their potential. Lessons learned with the plant viruses could also be applied to the new ssDNA viruses being found almost daily which infect fungi, algae and animals—and some of these or elements derived from them may even be found to be useful in plant systems, and vice versa.

Acknowledgments We gratefully acknowledge financial assistance from the University of Cape Town, the National Research Foundation and the South African AIDS Vaccine Initiative, Guy Regnard and James Dale for illustrations, and lab members past and present for their sterling work.

References

Aronson MN, Complainville A, Clerot D et al (2002) In planta protein–protein interactions assessed using a nanovirus-based replication and expression system. Plant J 31:767–775

Aronson MN, Meyer AD, Gyorgyey J, Katul L, Vetten HJ, Gronenborn B, Timchenko T (2000) *Clink*, a nanovirus-encoded protein, binds both pRB and SKP1. J Virol 74:2967–2972

Bottcher B, Unseld S, Ceulemans H, Russell RB, Jeske H (2004) Geminate structures of African cassava mosaic virus. J Virol 78:6758–6765

Boulton MI (1995) Agrobacterium-mediated transfer of geminiviruses to plant tissues. Methods Mol Biol 49:77–93

Briddon RW, Patil BL, Bagewadi B et al (2010) Distinct evolutionary histories of the DNA-A and DNA-B components of bipartite begomoviruses. BMC Evol Biol 10:97

Briddon RW, Stanley J (2006) Subviral agents associated with plant single-stranded DNA viruses. Virology 344:198–210

Briddon RW, Watts J, Markham PG et al (1989) The coat protein of beet curly top virus is essential for infectivity. Virology 172:628–633

Casado CG, Javier OG, Padron E, Bean SJ, McKenna R, Agbandje-McKenna M, Boulton MI (2004) Isolation and characterization of subgenomic DNAs encapsidated in “single” T = 1 isometric particles of maize streak virus. Virology 323:164–171

Collens JI, Mason HS, Curtis WR (2007) Agrobacterium-mediated viral vector-amplified transient gene expression in *Nicotiana glutinosa* plant tissue culture. Biotechnol Prog 23:570–576

Donson J, Gunn HV, Woolston CJ et al (1988) Agrobacterium-mediated infectivity of cloned digitaria streak virus DNA. Virology 162:248–250

Dry IB, Krake LR, Rigden JE, Rezaian MA (1997) A novel subviral agent associated with a geminivirus: the first report of a DNA satellite. Proc Natl Acad Sci USA 94:7088–7093

Dugdale B, Beetham PR, Becker DK et al (1998) Promoter activity associated with the intergenic regions of banana bunchy top virus DNA-1 to -6 in transgenic tobacco and banana cells. J Gen Virol 79(Pt 10):2301–2311

Fischer R, Vaquero-Martin C, Sack M et al (1999) Towards molecular farming in the future: transient protein expression in plants. Biotechnol Appl Biochem 30(Pt 2):113–116

Fofana IB, Sangare A, Collier R, Taylor C, Fauquet CM (2004) A geminivirus-induced gene silencing system for gene function validation in cassava. Plant Mol Biol 56:613–624

Frischmuth T, Ringel M, Kocher C (2001) The size of encapsidated single-stranded DNA determines the multiplicity of African cassava mosaic virus particles. J Gen Virol 82:673–676

Gilbertson RL, Sudarshana M, Jiang H et al (2003) Limitations on geminivirus genome size imposed by plasmodesmata and virus-encoded movement protein: insights into DNA trafficking. Plant Cell 15:2578–2591

Giritch A, Marillonnet S, Engler C et al (2006) Rapid high-yield expression of full-size IgG antibodies in plants coinfected with noncompeting viral vectors. Proc Natl Acad Sci USA 103:14701–14706

Gleba Y, Klimyuk V, Marillonnet S (2005) Magnifection—a new platform for expressing recombinant vaccines in plants. Vaccine 23:2042–2048

Gleba Y, Klimyuk V, Marillonnet S (2007) Viral vectors for the expression of proteins in plants. Curr Opin Biotechnol 18:134–141

Golenberg EM, Sather DN, Hancock LC et al (2009) Development of a gene silencing DNA vector derived from a broad host range geminivirus. Plant Methods 5:9

Grigoras I, Timchenko T, Katul L et al (2009) Reconstitution of authentic nanovirus from multiple cloned DNAs. J Virol 83:10778–10787

Grimsley N, Hohn B, Hohn T et al (1986) "Agroinfection," an alternative route for viral infection of plants by using the Ti plasmid. Proc Natl Acad Sci USA 83:3282–3286

Grimsley N, Hohn T, Davies JW et al (1987) *Agrobacterium*-mediated delivery of infectious maize streak virus into maize plants. Nature 325:177–179

Gronenborn B (2004) Nanoviruses: genome organisation and protein function. Vet Microbiol 98:103–109

Gutierrez C (1999) Geminivirus DNA replication. Cell Mol Life Sci 56:313–329

Gutierrez C (2000) DNA replication and cell cycle in plants: learning from geminiviruses. EMBO J 19:792–799

Gutierrez C, Ramirez-Parra E, Mar CM et al (2004) Geminivirus DNA replication and cell cycle interactions. Vet Microbiol 98:111–119

Halley-Stott RP, Tanzer F, Martin DP et al (2007) The complete nucleotide sequence of a mild strain of Bean yellow dwarf virus. Arch Virol 152:1237–1240

Hanley-Bowdoin L, Settlage SB, Orozco BM, Nagar S, Robertson D (2000) Geminiviruses: models for plant DNA replication, transcription, and cell cycle regulation. Crit Rev Biochem Mol Biol 35:105–140

Hanley-Bowdoin L, Settlage SB, Robertson D (2004) Reprogramming plant gene expression: a prerequisite to geminivirus DNA replication. Mol Plant Pathol 5:149–156

Harding RM, Burns TM, Dale JL (1991) Virus-like particles associated with banana bunchy top disease contain small single-stranded DNA. J Gen Virol 72(Pt 2):225–230

Hayes RJ, Coutts RH, Buck KW (1989) Stability and expression of bacterial genes in replicating geminivirus vectors in plants. Nucleic Acids Res 17:2391–2403

Hayes RJ, Petty IT, Coutts RH et al (1988) Gene amplification and expression in plants by a replicating geminivirus vector. Nature 334:179–182

Heath JD, Boulton MI, Raineri DM et al (1997) Discrete regions of the sensor protein virA determine the strain-specific ability of Agrobacterium to agroinfect maize. Mol Plant Microbe Interact 10:221–227

Hefferon KL, Fan Y (2004) Expression of a vaccine protein in a plant cell line using a geminivirus-based replicon system. Vaccine 23:404–410

Hefferon KL, Kipp P, Moon YS (2004) Expression and purification of heterologous proteins in plant tissue using a geminivirus vector system. J Mol Microbiol Biotechnol 7:109–114

Hiatt A, Cafferkey R, Bowdish K (1989) Production of antibodies in transgenic plants. Nature 342:76–78

Hormuzdi SG, Bisaro DM (1993) Genetic analysis of beet curly top virus: evidence for three virion sense genes involved in movement and regulation of single- and double-stranded DNA levels. Virology 193:900–909
Hormuzdi SG, Bisaro DM (1995) Genetic analysis of beet curly top virus: examination of the roles of L2 and L3 genes in viral pathogenesis. Virology 206:1044–1054
Huang Z, Chen Q, Hjelm B, Arntzen C, Mason H (2009) A DNA replicon system for rapid high-level production of virus-like particles in plants. Biotechnol Bioeng 103:706–714
Huang Z, Phoolcharoen W, Lai H, Piensook K, Cardineau G, Zeitlin L, Whaley KJ, Arntzen CJ, Mason HS, Chen Q (2010) High-level rapid production of full-size monoclonal antibodies in plants by a single-vector DNA replicon system. Biotechnol Bioeng 106:9–17
Jupin I, de Kouchkovsky F, Gronenborn B (1994) Movement of tomato yellow leaf curl geminivirus (TYLCV): involvement of the protein encoded by ORF C4. Virology 204:82–90
Kammann M, Matzeit V, Schmidt B, Schell J, Walden R, Gronenborn B (1991) Geminivirus-based shuttle vectors capable of replication in *Escherichia coli* and monocotyledonous plant cells. Gene 104:247–252
Kim KI, Sunter G, Bisaro DM, Chung IS (2007) Improved expression of recombinant GFP using a replicating vector based on Beet curly top virus in leaf-disks and infiltrated *Nicotiana benthamiana* leaves. Plant Mol Biol 64:103–112
Koonin EV, Ilyina TV (1992) Geminivirus replication proteins are related to prokaryotic plasmid rolling circle DNA replication initiator proteins. J Gen Virol 73(Pt 10):2763–2766
Kotlizky G, Boulton MI, Pitaksutheepong C, Davies JW, Epel BL (2000) Intracellular and intercellular movement of maize streak geminivirus V1 and V2 proteins transiently expressed as green fluorescent protein fusions. Virology 274:32–38
Lageix S, Catrice O, Deragon JM, Gronenborn B, Pelissier T, Ramirez BC (2007) The nanovirus-encoded *Clink* protein affects plant cell cycle regulation through interaction with the retinoblastoma-related protein. J Virol 81:4177–4185
Laufs J, Wirtz U, Kammann M, Matzeit V, Schaefer S, Schell J, Czernilofsky AP, Baker B, Gronenborn B (1990) Wheat dwarf virus Ac/Ds vectors: expression and excision of transposable elements introduced into various cereals by a viral replicon. Proc Natl Acad Sci USA 87:7752–7756
Lazarowitz SG, Beachy RN (1999) Viral movement proteins as probes for intracellular and intercellular trafficking in plants. The Plant Cell 11:535–548
Liu H, Boulton MI, Davies JW (1997) Maize streak virus coat protein binds single- and double-stranded DNA in vitro. J Gen Virol 78(Pt 6):1265–1270
Liu H, Boulton MI, Oparka KJ, Davies JW (2001) Interaction of the movement and coat proteins of Maize streak virus: implications for the transport of viral DNA. J Gen Virol 82:35–44
Liu H, Boulton MI, Thomas CL, Prior DA, Oparka KJ, Davies JW (1999) Maize streak virus coat protein is karyophyllic and facilitates nuclear transport of viral DNA. Mol Plant Microbe Interact 12:894–900
Maclean J, Koekemoer M, Olivier AJ, Stewart D, Hitzeroth II, Rademacher T, Fischer R, Williamson AL, Rybicki EP (2007) Optimization of human papillomavirus type 16 (HPV-16) L1 expression in plants: comparison of the suitability of different HPV-16 L1 gene variants and different cell-compartment localization. J Gen Virol 88:1460–1469
Mansoor S, Khan SH, Bashir A, Saeed M, Zafar Y, Malik KA, Briddon R, Stanley J, Markham PG (1999) Identification of a novel circular single-stranded DNA associated with cotton leaf curl disease in Pakistan. Virology 259:190–199
Matzeit V, Schaefer S, Kammann M, Schalk HJ, Schell J, Gronenborn B (1991) Wheat dwarf virus vectors replicate and express foreign genes in cells of monocotyledonous plants. Plant Cell 3:247–258
Meyers A, Chakauya E, Shephard E, Tanzer FL, Maclean J, Lynch A, Williamson AL, Rybicki EP (2008) Expression of HIV-1 antigens in plants as potential subunit vaccines. BMC Biotechnol 8:53
Mor TS, Moon YS, Palmer KE, Mason HS (2003) Geminivirus vectors for high-level expression of foreign proteins in plant cells. Biotechnol Bioeng 81:430–437

Muangsan N, Robertson D (2004) Geminivirus vectors for transient gene silencing in plants. Methods Mol Biol 265:101–115
Nawaz-ul-Rehman MS, Fauquet CM (2009) Evolution of geminiviruses and their satellites. FEBS Lett 583:1825–1832
Needham PD, Atkinson RG, Morris BAM, Gardner RC, Gleave AP (1998) GUS expression patterns from a tobacco yellow dwarf virus-based episomal vector. Plant Cell Rep 17:631–639
Nishigawa H, Miyata S, Oshima K, Sawayanagi T, Komoto A, Kuboyama T, Matsuda I, Tsuchizaki T, Namba S (2001) In planta expression of a protein encoded by the extrachromosomal DNA of a phytoplasma and related to geminivirus replication proteins. Microbiology 147:507–513
Noueiry AO, Lucas WJ, Gilbertson RL (1994) Two proteins of a plant DNA virus coordinate nuclear and plasmodesmal transport. Cell 76:925–932
Padidam M, Beachy RN, Fauquet CM (1996) The role of AV2 ("precoat") and coat protein in viral replication and movement in tomato leaf curl geminivirus. Virology 224:390–404
Palmer KE, Rybicki EP (1997) The use of geminiviruses in biotechnology and plant molecular biology, with particular focus on mastreviruses. Plant Sci 129:115–130
Palmer KE, Rybicki EP (1998) The molecular biology of mastreviruses. Adv Virus Res 50:183–234
Palmer KE, Rybicki EP (2001) Investigation of the potential of maize streak virus to act as an infectious gene vector in maize plants. Arch Virol 146:1089–1104
Palmer KE, Schnippenkoetter WH, Rybicki EP (1998) Geminivirus isolation and DNA extraction. Methods Mol Biol 81:41–52
Palmer KE, Thomson JA, Rybicki EP (1999) Generation of maize cell lines containing autonomously replicating maize streak virus-based gene vectors. Arch Virol 144:1345–1360
Pascal E, Goodlove PE, Wu LC, Lazarowitz SG (1993) Transgenic tobacco plants expressing the geminivirus BL1 protein exhibit symptoms of viral disease. Plant Cell 5:795–807
Pascal E, Sanderfoot AA, Ward BM, Medville R, Turgeon R, Lazarowitz SG (1994) The geminivirus BR1 movement protein binds single-stranded DNA and localizes to the cell nucleus. Plant Cell 6:995–1006
Paszkowski U, Zhang SB, Potrykus I, Paszkowski J (1993) Replication of the DNA A component of African cassava mosaic virus in a heterologous system. J Gen Virol 74(Pt 12):2725–2729
Qin S, Ward BM, Lazarowitz SG (1998) The bipartite geminivirus coat protein aids BR1 function in viral movement by affecting the accumulation of viral single-stranded DNA. J Virol 72:9247–9256
Raghavan V, Malik PS, Choudhury NR, Mukherjee SK (2004) The DNA-A component of a plant geminivirus (Indian mung bean yellow mosaic virus) replicates in budding yeast cells. J Virol 78:2405–2413
Reavy B, Bagirova S, Chichkova NV, Fedoseeva SV, Kim SH, Vartapetian AB, Taliansky ME (2007) Caspase-resistant VirD2 protein provides enhanced gene delivery and expression in plants. Plant Cell Rep 26:1215–1219
Regnard GL, Halley-Stott RP, Tanzer FL, Hitzeroth II, Rybicki EP (2010) High level protein expression in plants through the use of a novel autonomously replicating geminivirus shuttle vector. Plant Biotechnol J 8:38–46
Rekab D, Carraro L, Schneider B, Seemuller E, Chen J, Chang CJ, Locci R, Firrao G (1999) Geminivirus-related extrachromosomal DNAs of the X-clade phytoplasmas share high sequence similarity. Microbiology 145(Pt 6):1453–1459
Rochester DE, Kositratana W, Beachy RN (1990) Systemic movement and symptom production following agroinoculation with a single DNA of tomato yellow leaf curl geminivirus (Thailand). Virology 178:520–526
Rodriguez-Negrete EA, Carrillo-Tripp J, Rivera-Bustamante RF (2009) RNA silencing against geminivirus: complementary action of posttranscriptional gene silencing and transcriptional gene silencing in host recovery. J Virol 83:1332–1340
Rojas MR, Noueiry AO, Lucas WJ, Gilbertson RL (1998) Bean Dwarf mosaic geminivirus movement proteins recognize DNA in a form- and size-specific manner. Cell 95:105–113

Rosario K, Duffy S, Breitbart M (2009) Diverse circovirus-like genome architectures revealed by environmental metagenomics. J Gen Virol 90:2418–2424
Rybicki EP (2009a) Plant-produced vaccines: promise and reality. Drug Discov Today 14:16–24
Rybicki EP (2009b) Third international conference on plant-based vaccines and antibodies. Expert Rev Vaccines 8:1151–1155
Sainsbury F, Lomonossoff GP (2008) Extremely high-level and rapid transient protein production in plants without the use of viral replication. Plant Physiol 148:1212–1218
Saitou N, Nei M (1987) The neighbor-joining method: a new method for reconstructing phylogenetic trees. Mol Biol Evol 4:406–425
Saunders K, Bedford ID, Briddon RW, Markham PG, Wong SM, Stanley J (2000) A unique virus complex causes Ageratum yellow vein disease. Proc Natl Acad Sci USA 97:6890–6895
Saunders K, Stanley J (1999) A nanovirus-like DNA component associated with yellow vein disease of Ageratum conyzoides: evidence for interfamilial recombination between plant DNA viruses. Virology 264:142–152
Schob H, Kunz C, Meins F Jr (1997) Silencing of transgenes introduced into leaves by agroinfiltration: a simple, rapid method for investigating sequence requirements for gene silencing. Mol Gen Genet 256:581–585
Selth LA, Randles JW, Rezaian MA (2002) *Agrobacterium tumefaciens* supports DNA replication of diverse geminivirus types. FEBS Lett 516:179–182
Shen WH, Hohn B (1992) Excision of a transposable element from a viral vector introduced into maize plants by agroinfection. Plant J 2:35–42
Shen WH, Hohn B (1994) Amplification and expression of the beta-glucuronidase gene in maize plants by vectors based on maize streak virus. Plant J 5:227–236
Shepherd DN, Martin DP, van der Walt E, Dent K, Varsani A, Rybicki EP (2010) Maize streak virus: an old and complex 'emerging' pathogen. Mol Plant Pathol 11:1–12
Stanley J (1993) Geminiviruses: plant viral vectors. Curr Opin Genet Dev 3:91–96
Stanley J, Bisaro DM, Briddon RW, Brown JK, Fauquet CM, Harrison BD, Rybicki EP, Stenger DC (2005) Geminiviridae. In: Fauquet CM, Mayo MA, Maniloff J, Desselberger U, Ball LA (eds) International Committee on the Taxonomy of Viruses (VIIIth Report). Elsevier/Academic Press, London, pp 301–326
Stanley J, Latham JR, Pinner MS, Bedford I, Markham PG (1992) Mutational analysis of the monopartite geminivirus beet curly top virus. Virology 191:396–405
Sugimoto K, Otsuki Y, Saji S, Hirochika H (1994) Transposition of the maize Ds element from a viral vector to the rice genome. Plant J 5:863–871
Takamatsu N, Ishikawa M, Meshi T, Okada Y (1987) Expression of bacterial chloramphenicol acetyltransferase gene in tobacco plants mediated by TMV-RNA. EMBO J 6:307–311
Tamilselvi D, Anand G, Swarup S (2004) A geminivirus AYVV-derived shuttle vector for tobacco BY2 cells. Plant Cell Rep 23:81–90
Thomas JE, Dietzgen RG (1991) Purification, characterization and serological detection of virus-like particles associated with banana bunchy top disease in Australia. J Gen Virol 72(Pt 2): 217–224
Thomas JE et al (2010) Two novel mastreviruses from chickpea (Cicer arietinum) in Australia. Arch Virol 155(11):1777–1788 (Epub 2010 Aug 24)
Timchenko T, Katul L, Aronson M, Vega-Arreguin JC, Ramirez BC, Vetten HJ, Gronenborn B (2006) Infectivity of nanovirus DNAs: induction of disease by cloned genome components of Faba bean necrotic yellows virus. J GenVirol 87:1735–1743
Timmermans MCP, Das OP, Messing J (1994) Geminiviruses and their uses as extrachromosomal replicons. Annu Rev Plant Physiol Plant Mol Biol 45:79–112
Turnage MA, Muangsan N, Peele CG, Robertson D (2002) Geminivirus-based vectors for gene silencing in Arabidopsis. Plant J 30:107–114
Vetten HJ, Chu PWG, Dale JL, Harding R, Hu J, Katul L, Kojima M, Randles JW, Sano Y, Thomas JE (2005) *Nanoviridae* In: Fauquet CM, Mayo MA, Maniloff J, Desselberger U, Ball LA (eds) Virus taxonomy, 8th report of the international committee on taxonomy of viruses, Elsevier/Academic Press, p 1259

von Arnim A, Stanley J (1992) Inhibition of African cassava mosaic virus systemic infection by a movement protein from the related geminivirus tomato golden mosaic virus. Virology 187:555–564

Wanitchakorn R, Hafner GJ, Harding RM, Dale JL (2000) Functional analysis of proteins encoded by banana bunchy top virus DNA-4 to -6. J Gen Virol 81:299–306

Ward A, Etessami P, Stanley J (1988) Expression of a bacterial gene in plants mediated by infectious geminivirus DNA. EMBO J 7:1583–1587

Wartig L, Kheyr-Pour A, Noris E, de Kouchovsky F, Jouanneau F, Gronenborn B, Jupin I (1997) Geneticanalysis of the monopartite tomato yellow leaf curl geminivirus: roles of V1,V2, and C2 ORFs in viral pathogenesis. Virology 228:132–140

Wirtz U, Osborne B, Baker B (1997) Ds excision from extrachromosomal geminivirus vector DNA is coupled to vector DNA replication in maize. Plant J 11:125–135

Wu CY, Yang SH, Lai YC, Lin NS, Hsu YH, Hu CC (2007) Unit-length, single-stranded circular DNAs of both polarity of begomoviruses are generated in *Escherichia coli* harboring phage M13-cloned begomovirus genome with single copy of replication origin. Virus Res 125:14–28

Yu X, Li B, Fu Y, Jiang D, Ghabrial SA, Li G, Peng Y, Xie J, Cheng J, Huang J, Yi X (2010) A geminivirus-related DNA mycovirus that confers hypovirulence to a plant pathogenic fungus. Proc Natl Acad Sci USA 107:8387–8392

Yusibov V, Rabindran S, Commandeur U, Twyman RM, Fischer R (2006) The potential of plant virus vectors for vaccine production. Drugs RD 7:203–217

Zhang W, Olson NH, Baker TS, Faulkner L, gbandje-McKenna M, Boulton MI, Davies JW, McKenna R (2001) Structure of the Maize streak virus geminate particle. Virology 279:471–477

Zhang X, Mason H (2006) Bean Yellow Dwarf Virus replicons for high-level transgene expression in transgenic plants and cell cultures. Biotechnol Bioeng 93:271–279

Plant Viral Epitope Display Systems for Vaccine Development

Denis Leclerc

Abstract The 'easiest' vaccines, base on production of neutralizing antibodies, have been made. With the emergence of chronic diseases, vaccine developers have understood the importance to trigger an efficient cellular mediated immune response (CTL response) to respond to this medical need. Several options are currently in development and the utilization of plant virus as vaccine platform for the trigger of a CTL response is considered as an interesting avenue. The highly ordered structures of plant viruses are good triggers of the innate immune system, which in turn, is used to initiate an immune response to a vaccine target. It is likely that plant viruses will play an important role in the development of the vaccine of the futures even if there is still several challenges to face.

Abbreviations

APCs	Antigen presentation cells
CD8+	Cytotoxic T cells with CD8 surface protein are called *CD8+* T cells
CP	Coat protein
CPMV	Cowpea mosaic virus
DCs	Dendritic cells
CTL	Cytotoxic T lymphocytes
FMDV	Foot-and-mouth disease
Gp100	Melanosomal matrix protein whose expression is closely correlated with cellular melanin content
GFP	Gellyfish fluorescent protein

D. Leclerc (✉)
Infectious Diseases Research Centre, Pavilion CHUL, Laval University, 2705 Boulevard Laurier, Quebec City, PQ G1V 4G2, Canada
e-mail: Denis.Leclerc@crchul.ulaval.ca

Current Topics in Microbiology and Immunology (2011) 375: 47–59

DOI: 10.1007/82_2011_183

Published Online: 22 October 2011

HBV	Hepatitis B virus
HCV	Hepatitis C virus
HIV-1	Human immunodeficiency virus-1
HLA	Human leucocyte antigene
HPV	Human papilloma virus
IFN-γ	Interferon gamma
LCMV	Lymphocytic choriomeningitis virus
MaMV	Malva mosaic virus
MHC class 1	One of two primary classes of major histocompatibility complex (MHC) molecules I
PAMPs	Pathogen associated molecular patterns
PapMV	Papaya mosaic virus
PPV	Procine parvorvirus
PVX	Potato virus X
TAP	Transporter associated with antigen processing
TMV	Tobacco mosaic virus
VLPs	Virus like particles

Contents

1 Introduction

Vaccination has proved the most efficient method in human efforts to control infectious disease, leading to the eradication of small pox; the near eradication of polio; and excellent control of measles, rubella, mumps, pertussis and diphtheria (Plotkin 2005). The practice of vaccination has been highlighted as one of the major causes of the increase in life expectancy, and vaccines still have a role to play in further improving the human condition (Boraschi et al. 2010).

1.1 Vaccination in the Elderly

Despite the obvious success of vaccines, they usually provide only partial protection. The case of the 'flu' vaccine illustrates this point well since the clinical effectiveness of flu vaccines in older adults (aged $\geq$65 years) typically ranges from 30 to 40% (Palache 1997; Strassburg et al. 1986). Consequently, flu infection induces high morbidity in the elderly (Deng et al. 2004; Simonsen et al. 2005). A large proportion of this population remains susceptible to infection, with the additional risk of developing complications, despite the seasonal influenza vaccination (Simonsen et al. 2007; Goodwin et al. 2006). This is a major issue given the aging population in developed countries. More immunogenic vaccines in the elderly could decrease significantly the numbers of hospital visits by people of this age group, and consequently decrease the costs of medical care. In countries like Canada, that depend on universal health programs, an increase of the cost with an aging population in the coming years is foreseen, which has recently become of major concern to political leaders.

1.2 Vaccines Inducing CTL Response: A Challenge

Most vaccines currently on the market are known to induce the production of antibodies that neutralize the target pathogen, thus preventing infection of host cells. This approach, although valuable, is not sufficient to combat chronic diseases because antibodies alone cannot control the replication of those pathogens (O'Connell et al. 2009). Several viruses, which induce chronic diseases, including human immunodeficiency virus 1 (HIV-1) (Schief et al. 2009; Vaine et al. 2009) and hepatitis C virus (HCV) (Zeisel et al. 2007; Stamataki et al. 2008), or viruses causing pandemics like influenza, are known to mutate very fast and can thus escape neutralization by antibodies (Li and Poon 2009). It is well recognized that triggering a cellular mediated immune response (CTL) with vaccines designed towards conserved CTL epitopes is a preferred approach in the control of replication of such pathogens (Guidotti and Chisari 2009). The conserved epitopes targetted in such cases are usually located in viral proteins, like viral nucleocapsids of enveloped virus or replicases, that are not subject to neutralization by antibodies because they are found in the interior of the virus and are produced only inside the infected host cells (Doherty et al. 2006). CTL epitopes are presented by the class I major histocompatibility complex (MHC) on the surface of infected cells, which leads to activation of CD8+ specific T-cells directed towards those peptides, and subsequent elimination of the infected cells. Vaccination, which improves the CTL response, increases the pool of specific CD8+ cells directed to specific CTL epitopes of the virus, which, in return, allows the infection to be controlled. This is a new concept in vaccination because it does not provide a sterilizing immunity through neutralizing antibodies. With this approach, the patient will still become infected but will not suffer the symptoms of infection because the pathogen will be rapidly controlled by the CD8+ mediated cellular response that was stimulated by the prophylactic vaccination.

1.3 Immunogenicity of Recombinant Subunit Vaccines

Vaccines currently under development are composed of synthetic, recombinant, or highly purified subunit antigens. In order to make them safer than whole-inactivated or live-attenuated vaccines, subunit vaccines are designed to include only those antigens required for protective immunization. However, the purity of the subunit antigens and the absence of the self-adjuvanting immunomodulatory components associated with traditional vaccines often results in weaker immunogenicity. Formulation of subunit vaccines with a potent adjuvant (a component added to vaccines to make them more immunogenic) is therefore an attractive approach to improving their performance.

Ideally, we wish to increase the immunogenicity of the vaccine in a two-pronged offensive, targeting both humoral (antibodies) and the cytotoxic response (CTL), and thus offer a broad and long lasting immunity against pathogens. From the above discussion it is clear that the use of adjuvants and vaccine platforms is urgently needed to improve the immunogenicity of vaccines in older people, to combat the threat of pandemics, and to fight chronic viral diseases. Alum—the most widely accepted adjuvant in the US—has been used since 1926 but is unable to induce a cellular response (Atmar and Keitel 2009). Consequently, innovative and efficient new technologies are needed.

2 Plant Viruses as a Vaccine Platform

A new and emerging class of molecules known as virus like particles (VLPs) has triggered much interest lately. VLPs have been used successfully in human papilloma virus (HPV) (Harper et al. 2004), and hepatitis B virus (HBV) (Fagan et al. 1987) vaccines, and were showed to be very promising in vaccination against influenza (Pushko et al. 2007). VLPs, which are made of the structural proteins of the viral pathogen, can be produced in powerful expression systems like yeast (Fagan et al. 1987; Harper et al. 2004) or insect (Pushko et al. 2007) cells, and provide protection against the pathogen from which they are derived. They are immunogenic but are usually used together with Alum as an adjuvant.

Plant viruses, which are not pathogenic to humans, are structurally similar to VLPs and show strong potential for vaccine development. Plant viruses are made up of several coat protein (CP) subunits assembled around a viral genome, usually an RNA genome, in a highly repetitive and crystalline manner that is recognized by the immune system as a pathogen associated molecular pattern (PAMP) (Denis et al. 2008; Lacasse et al. 2008; Smith et al. 2009; Gonzales et al. 2009; Savard et al. 2011). They are phylogenetically distant from the mammalian immune system, which contributes to their immunogenicity. Plant viruses have been shown to trigger a humoral response to peptides fused directly to their surface in several systems, including tobacco mosaic virus (TMV) (Smith et al. 2009), cowpea mosaic virus (CPMV) (Liu et al. 2005), alfalfa mosaic virus (AlMV) (Yusibov et al. 2005), papaya mosaic virus (PapMV) (Denis et al. 2007, 2008) and several others (Cañizares et al. 2005). Besides PapMV, from which VLPs were produced from expression of recombinant CP in *Escherichia coli*, all other plant viruses tested in vaccination have been produced and purified from plants infected with a recombinant chimeric plant virus. The expression and purification of viruses in plants is attractive because growing plants in the field or in the greenhouse requires only a minimum of equipment and is possible at low cost (Streatfield and Howard 2003; Smith et al. 2009). Plant viruses can replicate to very high titres *in planta*, providing an efficient method of production. It was estimated that between 1.5 and 2.5 kg of purified TMV or recombinant TMV can be recovered per hectare of infected plants (Smith et al. 2009), which corresponds to approximately 5 million doses of 250 μg each. However, the major limitation of this approach is that the plant viral CP, which is the major component of plant viruses, is often involved in important aspects of the viral life cycle, such as systemic movement of the virus through the host plant (Cruz et al. 1996; Lico et al. 2006), replication of the virus *in planta* (Bol 2005), or in the assembly of the virus particle (Tremblay et al. 2006). If fusion of an antigenic peptide interferes with one of those critical functions, the yield of chimeric virus can be decreased significantly (Bendahmane et al. 1999). In contrast, expression of a plant virus CP in bacteria does not rely on viral replication and has a higher chance of generating a product if fusion of the peptide is tolerated and permits VLP formation (Tremblay et al. 2006; Denis et al. 2007, 2008). Finally, in contrast to purified plant viruses, plant-virus-based VLPs produced in bacteria do not contain the viral genome, which could avoid the safety concerns of regulatory bodies. Therefore, depending of the product and the market that is targeted, the use of plant or bacterial systems have their own advantages.

3 Plant Viruses and Induction of the CTL Response

One of the most remarkable properties of plant-virus-based VLPs is their ability to trigger a CTL response. This property has been shown for several rod-shaped plant viruses including TMV (McCormick et al. 2006), PapMV (Leclerc et al. 2007; Lacasse et al. 2008) and potato virus X (PVX) (Lico et al. 2009). Chimeric VLPs derived from the nucleocapsid of HPV (Liu et al. 2002) and porcine parvovirus (PPV) (Sedlik et al. 1997), which are icosahedric viruses, were shown to trigger a CTL response to an epitope fused to their structure. It is likely that icosahedral plant viruses are also able to trigger CTL but, so far, demonstration of this property has been limited to rod-shaped viruses.

To trigger a CTL response, the VLPs must be taken up via endocytosis by antigen presentation cells (APCs), thereby reaching the endosomal compartment where they are digested by proteases into peptides that can be presented by MHC class I. The size of plant viruses, which ranges between 50 and 500 nm in length or diameter, is optimal to trigger this reaction in dedicated APCs such as dendritic cells (DCs) (Fifis et al. 2004; Gonzales et al. 2009). This process is called cross presentation of CTL epitopes. CTL epitopes usually arise from proteins that are derived from the cytoplasm of the cell. Plant viruses fused to a CTL epitope are taken up from the exterior of the cell before reaching the endosomal compartment. It is likely that animal immune systems have evolved to recognize, in a general manner, plant virus structure as a danger signal, which triggers their uptake and degradation by the cellular machinery. Cross-presentation of CTL epitopes from particulate structures like plant viruses occurs with an efficiency comparable to that of presentation on the MHC class II molecule, which specializes in presentation of external soluble antigens (Harding and Song 1994; Storni and Bachmann 2004; Leclerc et al. 2007; Hanafi et al. 2010)—a property making them ideal for the development of efficient vaccines.

PapMV VLPs produced in bacteria have been shown to trigger cross presentation of CTL epitopes using the Lymphocytic choriomeningitis virus (LCMV) model in mice (Lacasse et al. 2008). DCs pulsed with PapMV-like particles displaying the LCMV p33 immuno-dominant CTL epitope (PapMV-p33) were efficiently processed and were able to present the viral epitope to p33-specific CD8+ T cells. Immunization of mice in the absence of adjuvant with PapMV-p33 develop p33-specific effector CTLs that expanded rapidly following LCMV challenge, which allowed protection of mice vaccinated against LCMV infection in a dose-dependent manner (Fig. 1) (Lacasse et al. 2008). In contrast to other VLPs described so far, PapMV-like particles directly prime protective CTL responses in vivo without the help of an adjuvant (i.e., CpGs or anti-CD40). This ability probably resides in their intrinsic capacity to induce DC maturation. McCormick et al. (2006) reported that recombinant chimeric TMV particles also induce DC maturation, demonstrating that recognition of plant virus PAMPs by DCs is not restricted to PapMV but is probably a more general phenomenon. In another system, the presence of chimeric TMV alone was not sufficient to induce DC maturation; the use CpG DNA as an adjuvant was needed to confer partial protection against tumor challenge (McCormick et al. 2006). It is unclear at this point if PapMV VLPs are more effective than TMV, HBV or other VLPs since they have not been compared within the same experiment. Nevertheless, the examples cited above show the high potential of plant viruses or plant-virus-based VLPs to induce an efficient CTL response.

In another study, we also showed that PapMV-derived VLP can mediate MHC class I cross-presentation in human APCs in a proteasome-independent manner as shown by the lack of inhibition of MHC class I loading using lactocystein and MG-132 proteosome inhibitors (Leclerc et al. 2007). PapMV VLPs completely bypass the transporter associated with the antigen (TAP) system that is normally used for loading MHC class I molecules. Chimeric PapMV VLPs harbouring an influenza M1 epitope or the melanoma gp100 CTL epitope pulsed to human APCs (CD40 activated B cells and DCs) efficiently expanded M1 and gp100

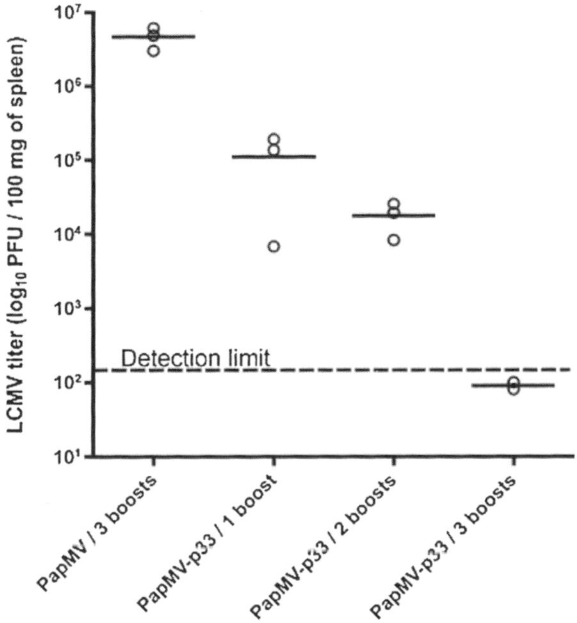

Fig. 1 Immunization with PapMV-p33 VLPs induces protective immunity against LCMV infection. Mice were injected i.v. with 100 μg of VLPs in PBS, followed by identical recall injections 10, 20, and 30 days later. Seven days after the injections, immunized mice were infected with 200 PFU of LCMV. LCMV titers were determined 5 days after infection (taken from Lacasse et al. 2008)

HLA-A*0201 epitope-specific T cells. Expanded T cells seem to be highly reactive, and avidly produce high levels of IFN-γ secretion at very low peptide concentrations (Leclerc et al. 2007). Cross-presentation mediated by the HBV VLP with CpG motifs for delivering a MHC class I epitope from LCMV has also been shown to be TAP independent (Ruedl et al. 2002). PapMV VLPs, with their complex repetitive structures, may act similarly, and we are currently investigating the exact processing mechanism involved in epitope release. Two recent reports showed that engineered PVX or malva mosaic virus (MaMV), two potexviruses related to PapMV, fused to a CTL epitope of influenza A virus, were able to trigger a CTL response to this epitope as shown by an enzyme linked immunospot (ELISPOT) assay in a mouse system (Lico et al. 2009) or in human cells (Hanafi et al. 2010), which is consistent with previous results obtained with the PapMV platform suggesting that this property is not unique to PapMV and is shared by other members of this plant virus family or viruses sharing a similar structure.

4 Conjugation With Large Antigens

4.1 Genetic Fusion With CP

Despite the high potential of plant viruses or plant-derived VLPs for the development of a CTL response in vaccines, most approaches used so far are based on the use of only one CTL epitope that is restricted to one or a few types of HLA in patients. To render this approach universal, each fusion would need to be adapted to the HLA of each patient, which is obviously impractical for prophylactic vaccines but could potentially be considered for therapeutic vaccines against cancer. Ideally, the prophylactic vaccine should provide several epitopes able to cover the HLAs of the entire population. Therefore, virus engineering will need to be improved to accommodate the fusion of a full-length antigen or a large fragment of it. Three main approaches have been proposed to satisfy this requirement. The first is through the genetic fusion of the protein directly to the open reading frame of the CP. In most cases, the fusion of a large protein directly to the plant virus CP will lead to a chimeric protein that is unable to assemble into virus particles in an organized manner and, therefore, will be non immunogenic (Denis et al. 2007, 2008). To overcome this limitation, the use of the TMV 130-kDa protein read-through motif that determines the suppression of the an amber stop codon was used between the C-terminal end of the TMV CP and the protein extension, allowing the generation of a mixed pool of free and fused CP subunits that can assemble into heterogenous subunits (Turpen et al. 1995). An epitope of 21 amino acids was fused at the surface of TMV using this approach. It is anticipated that longer fusions could potentially be tolerated, although this has not yet been validated. In a similar approach, a full-length green fluorescent protein (GFP; 27 kDa) was tolerated on the surface of PVX (Cruz et al. 1996). In this approach, the GFP open reading frame was fused at the N-terminus of the PVX CP separated by the foot-and-mouth disease virus (FMDV) 2A peptide, which corresponds to the cleavage site of the cellular protease. The design of the cleavage site reduced the efficacy of cleavage; the resulting partial cleavage generated a mixed population of WT PVX CP and CP harboring the fusion of the full length GFP. This permitted accumulation of hybrid PVX particles with a diameter larger than that of the WT virus. Fusion of neomycintransferase (31 kDa), chloramphenicol acetyltransferase (25.6 kDa) and β-galactosidase (8.5 kDa) has also been shown to be successful and to generate hybrid virus particles, making this approach very promising (Cruz et al. 1996). Unfortunately, none of these approaches has yet been tested in animals to assess their potential as candidate vaccines. It is uncertain at this point if the fusion of large polypeptides at the surface of the plant virus platform will interfere with the immunogenicity of the particles.

4.2 Conjugation by Chemical Cross-Linking

Conjugation using chemical cross-linking of the antigen directly on the plant virus is a second approach that seems very promising (McCormick et al. 2006; Smith et al. 2006). Using a heterobifunctional linker such as sulfo-LC-SPDP, peptides (McCormick et al. 2006) or full-length proteins (Smith et al. 2006) can be linked directly to the cross-linker, followed by specific coupling to the surface of the TMV platform on which a surface-exposed lysine was made available to accept the reactive peptide. To link the full length GFP-SA (GFP fused to streptavidin) to the virus surface, the TMV platform was modified with NHS-PEO$_4$-biotin, which reacted chemically with the exposed lysine residues and prepared the platform to receive the GFP-SA antigen (Smith et al. 2006). This approach is flexible, efficient and easy to use. The immunization of animals with this complex leads to development of a GFP-specific antibody response. The use of the TMV platform with unlinked GFP-SA was unable to raise an immune response to the antigen. The conjugation appeared to break the TMV rods into smaller rods but the resulting product was still sufficiently immunogenic to be used for vaccine development. It is unknown if this approach can be used to trigger a CTL response to the antigen, but it is anticipated that this will work. Finally, a similar approach using CPMV (Paebody 2003) was described and proposed as a commercial method for the development of vaccines to infectious disease and cancer. The advantage of chemical coupling of the antigen over genetic fusion to the virus is the absence of interference of the antigen on the folding of the CP or the antigen since they are produced and purified separately. It has been reported that the large scale coupling of the antigen to the platform can be accomplished in a reproducible, as well as cost effective, manner (Jennings and Bachmann 2008).

4.3 High Avidity VLPs

A third approach is to engineer the surface of the vaccine platform or VLP with a small peptide ligand that will generate a high avidity molecule upon its presentation at the surface of the VLP. Usually, the specific ligand is selected by phage display, and it is cloned at the surface of the virus or the VLPs. This concept was shown to be successful using as a platform PapMV VLPs engineered to bind specifically fungal spores (Morin et al. 2007). We have recently shown that this concept can be used to improve the immune response to a target antigen (Fig. 2). The significant increase in IFN-γ secretion by the splenocytes of immunized animals suggests that high avidity VLPs improved the CTL response toward the antigen. This approach appears very promising and we are currently exploiting it in the design of innovative CTL-based vaccines.

5 Concluding Remarks

The "easiest" vaccines—based on the use of inactivated or attenuated pathogens—have all been achieved; the biggest challenge now is the development of vaccines for chronic viral diseases. It is well accepted that triggering a CTL response to highly conserved proteins of viral pathogens is the most suitable approach to controlling these diseases. Plant viruses have successfully induced cross-presentation of CTL epitopes on MHC class I, and can trigger a CTL response, but those proofs of concept were made with only one epitope at a time,

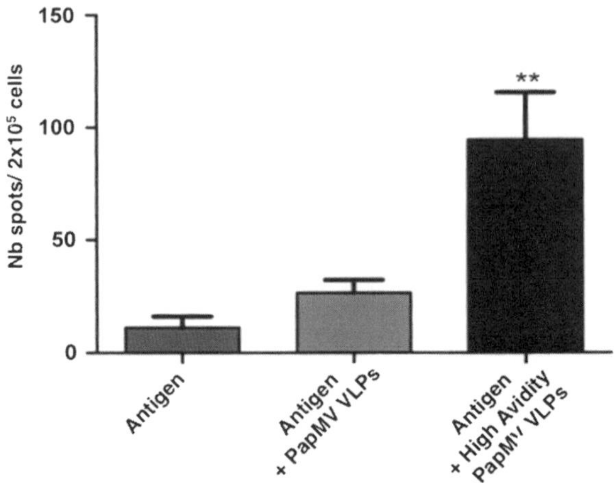

Fig. 2 High avidity PapMV VLPs improve the CTL response to the vaccine antigen. High avidity PapMV VLPs were engineered with a peptide specific to the antigen selected by phage display. The improvement in the adjuvant property of the high avidity VLPs was compared against empty PapMV VLPs with the antigen. Seven days after three injections given at 14 days interval, ELISPOT assyas were performed on isolated splenocytes reactivated with the antigen. Only high avidity PapMV VLPs improved significantly the secretion of IFN-γ as compared with the immunisation with the antigen alone or co-injected with empty PapMV VLP. ** $= P < 0.001$

which severely limits their utilization for general vaccination. Several approaches have been developed for conjugation of full-length antigen to plant virus, but more effort needs to be invested to prove efficacy in induction of CTL response to the antigen and in development of protection to infection. Also, it is important to consider that vaccines based on induction of CTL response can be efficient only if infection of the pathogen is allowed in the vaccinated patient since the CTL response eliminates only the reservoir of infection but does not neutralize the infection capacity of the invader. This is a new concept that differs considerably from the sterilizing immunity provided by antibodies. I believe it is safer at this stage to recommend the use of vaccines that trigger CTL as a complement to vaccines that provide sterilizing immunity through induction of an antibody response. The CTL complement can be used to provide additional protection to viral strains that escape neutralization by antibodies.

Another issue that must also be considered is the perception by regulatory bodies of plant-virus-based vaccines produced *in planta*. So far, no vaccine or protein product arising from a plant-based expression system has been licensed for immunization in humans. Therefore, there are several concerns related to potential contaminants that could follow such preparations and lead to toxicity. Prophylactic vaccines are used to prevent infection in healthy people, therefore, the risk has to be minimal to allow licensing of a new product. The cost of demonstrating minimal risk is extremely high since extensive clinical trials are necessary to determine the necessary proof of concept. The easiest avenue for plant virus vaccines in the mid-term will be the veterinarian market. This will help build confidence in these technologies and support their development for vaccines in humans at a later stage. An alternative to plants is the production of plant-virus-based VLPs in bacteria or yeast, which are expression systems already familiar to regulatory officers and the wider pharmaceutical industry in general. Our experience has shown that PapMV VLPs expressed in bacteria are as immunogenic as PapMV purified from infected plants. The expression of the CP of other plant viruses in bacteria were also shown to lead to self assembly of VLPs (Zayakina et al. 2009; Choudhury et al. 2009; Bragard et al. 2000; Yusibov et al. 1996; Zhao et al. 1995; Jagadish et al. 1991, 1993; Hwang et al. 1994) and this expression system is probably suitable for the CPs of several plant viruses.

Finally, the biggest challenge facing plant virus or plant-virus-based VLPs will be to show a competitive advantage over all the other emerging VLP systems based on the nucleocapsids of animal viruses or bacteriophages, which have also been shown to be valuable in vaccine development (Jennings and Bachmann 2008). It is likely that the VLPs that prove to easiest to engineer and that can accommodate conjugation with large antigens while showing the lowest toxicity will have an advantage over all other molecules of this class.

References

Atmar RL, Keitel WA (2009) Adjuvants for pandemic influenza vaccines. Curr Top Microbiol Immunol 333:323–44

Bendahmane M, Koo M, Karrer E, Beachy RN (1999) Display of epitopes on the surface of tobacco mosaic virus: impact of charge and isoelectric point of the epitope on virus–host interactions. J Mol Biol 290:9–20

Bol JF (2005) Replication of alfamo- and ilarmoviruses: role of the coat protein. Annu Rev Phytopathol. 43:39–62

Boraschi D, Del Giudice G, Dutel C et al (2010) Ageing and immunity: adressing immune senescence to ensure healthy ageing. Vaccine 28:3627–3631

Bragard C, Dincan GH, Wesley SV et al (2000) Virus like particles assemble in plants and bacteria expressing the coat protein gene of Indian peanut clump virus. J Gen Virol 81:267–272

Cañizares MC, Nicholson L, Lomonossof GP (2005) Use of viral vectors for vaccine production in plants. Immunol Cell Biol 83:263–270

Choudhury S, Kakkar V, Suman P et al (2009) Immunogenicity of zona pellucida glycoprotein-3 and spermatozoa YLP12 peptides presented on Johnson grass mosaic virus-like particles. Vaccine 27:2948–2953

Cruz SS, Chapman S, Roberts AG et al (1996) Assembly and movement of a plant virus carrying a green fluorescent protein overcoat. Proc Natl Acad Sci USA 93:6286–6290

Deng Y, Jing Y, Campbell AE, Graveinstein S (2004) Age-related impaired type 1 T cell responses to influenza: reduced activation ex vivo, decreased expansion in CTL culture in vitro, and blunted response to influenza vaccination in vivo in the elderly. J Immunol 172:3437–3446

Denis J, Majeau N, Acosta-Ramirez E, Savard C et al (2007) Immunogenicity of papaya mosaic virus like particlesfused to a hepatitis C virus epitope: evidence for the critical function of multimerization. Virology 363:59–68

Denis J, Acosta-Ramirez E, Zhao Y et al (2008) Development of a universal influenza a vaccine based on the M2e peptide fused to the papaya mosaic virus (PapMV) vaccine platform. Vaccine 26:3395–3403

Doherty PC, Turner SJ, Webby RG et al (2006) Influenza and the challenge for immunology. Nat Immunol 5:450–455
Fagan EA, Tolley P, Smith HM et al (1987) Hepatitis B vaccine: immunogenicity and follow-up including two year booster doses in high-risk health care personnel in a London teaching hospital. J Med Virol 21:49–56
Fifis T, Gamvrellis A, Crimeen-Irwin B et al (2004) Size-dependant immunogenicity: therapeutic and protective properties of nano-vaccines against tumors. J Immunol 173:3148–3154
Gonzales MJ, Plummer EM, Rae CS, Manchester M (2009) Interaction of cowpea mosaic virus (CPMV) nanoparticles with antigen presenting cells in vitro and in vivo. Plos One 4:e7981
Goodwin K, Viboud C, Simonsen L (2006) Antibody response to influenza vaccination in the elderly: a quantitative review. Vaccine 24:1159–1169
Guidotti LG, Chisari FV (2009) Immunobiology and pathogenesis of viral hepatitis. Annu Rev Pathol 1:23–61
Hanafi L-A, Bolduc M, Laliberté-Gagné M-E et al (2010) Two distinct chimeric potexviruses share antigenic cross-presentation properties of MHC class I epitopes. Vaccine 28:5617–5626
Harding CV, Song R (1994) Phagocytic processing of exogenous particulate antigens by macrophages for présentation by class I MHC molécule. J Immunol 153:4925–4933
Harper DM, Franco EL, Wheeler C et al (2004) Efficacy of a bivalent L1 virus-like particle vaccine in prevention of infection with human papillomavirus types 16 and 18 in young women: a randomised controlled trial. Lancet 364:1757–1765
Hwang DJ, Roberts IM, Wilson TMA (1994) Expression of tobacco mosaic virus coat protein and assembly of pseudovirus particles in Escherichia coli. Proc Nat Acad Sci USA 91:9067–9071
Jagadish MN, Ward CW, Gough KH et al (1991) Expression of potyvirus coat protein in Escherichia coli and yeast and its assembly into virus-like particles. J Gen Virol 72: 1543–1550
Jagadish MN, Huang D, Ward CW (1993) Site-directed mutagenesis of a potyvirus coat protein and its assembly in Escherichia coli. J Gen Virol 74:893–896
Jennings GT, Bachmann MF (2008) The coming age of virus-like paricle vaccines. Biol Chem 389:521–536
Lacasse P, Denis J, Lapointe R et al (2008) Novel plant virus-based vaccine induces protective CTL-mediated antiviral immunity through dendritic cell maturation. J Virol 82:785–794
Leclerc D, Beauseigle D, Denis J et al (2007) Proteasome-independent MHC class I cross-presentation mediated by papaya mosaic virus-like particles leads to the expansion of specific human T cells. J Virol 81:1319–1326
Li OT, Poon LL (2009) One step closer to universal influenza epitopes. Expert Rev Anti Infect Ther 7:687–690
Liu L, Cañizares MC, Monger W et al (2005) Cowpea moaic virus-based systems for the production of antigens and antibodies in plants. Vaccine 23:1788–1792
Liu XS, Liu WJ, Zhao KN et al (2002) Route of administration of chimeric BPV1 VLP determines the character of the induced immune responses. Immunol Cell Biol 80:21–29
Lico C, Capuano F, Renzone J et al (2006) Peptide display on potato vrus X: molecular features of the coat protein-fused peptide affeting cell-to-cell and phloem movement of chimeric virus particles. J Gen Virol 87:3103–3112
Lico C, Mancini C, Italiani P et al (2009) Plant-produced potato virus X chimeric particles displaying an influenza virus-derived paptide activate specific CD8+ T cells in mice. Vaccine 27:5069–5076
McCormick AA, Corbo TA, Wykoff-Clary S et al (2006) Chemical conjugate TMV-peptide bivalent fusion vaccines O, prove cellular immunity and tumor protection. Bioconjugate Chem 17:1330–1338
Morin H, Tremblay MH, Plante E et al (2007) High avidity binding of engineered papaya mosaic virus virus-like particles to resting spores of Plasmodiophora Brassicae. J Biotechnol 128:423–434
O'Connell KA, Bailey JR, Blankson JN (2009) Elucidating the élite: mechanisms of control of HIV-1 infection. Trends Pharmacol Sci 30:631–637

Palache AM (1997) Influenza vaccines: a reappraisal of their use. Drugs 54:841–856
Paebody DS (2003) A viral platform for chemical modification and multivalent display. J Nanobiotechnol 1:1–8
Plotkin SA (2005) Vaccines: past, present and future. Nat Med 11(Suppl 4):S5–S11
Pushko P, Tumpey TM, Van Hoeven N et al (2007) Evaluation of influenza virus-like particles and Novasome adjuvant as candidate vaccine for avian influenza. Vaccine 25:4283–4290
Ruedl C, Storni T, Lechner F et al (2002) Cross-presentation of virus-like particles by skin-derived CD8(+) dendritic cells: a dispensable role for TAP. Eur J Immunol 32:818–825
Savard C, Guérin A, Drouin K et al (2011) Improvement of the trivalent inactivated flu vaccine using PapMV nanoparticles. Plos One 6:e21522
Schief WR, Ban YE, Stamatatos L (2009) Challenges for structure-based HIV vaccine design. Curr Opin HIV AIDS 4:431–440
Sedlik C, Saron MF, Sarraseca J et al (1997) Recombinant parvovirus-like articles as an antigen carrier: a novel nonreplicative exogenous antigen to elicit protective antiviral cytotoxic t cells. Proc Natl Acad Sci USA 94:7503–7508
Simonsen L, Reichert TA, Viboud C et al (2005) Impact of influenza vaccination on seasonal mortality in the US elderly population. Arch Intern Med 165:265–272
Simonsen L, Taylor RJ, Viboud C et al (2007) Mortality benefits of influenza vaccination in elderly people: an ongoing controversy. Lancet Infect Dis 7:658–666
Smith ML, Lindbo JA, Dillard-Telm S et al (2006) Modified tobacco mosaic virus particles as scaffolds for display of protein antigens for vaccine applications. Virology 348:475–488
Smith ML, Fitzmaurice WP, Turpen TH, Palmer KE (2009) Display of peptides on the surface of tobacco mosaic virus particles. Curr Top Microbiol Immunol 332:13–31
Stamataki Z, Grove J, Balfe P et al (2008) Hepatitis C virus entry and neutralization. Clin Liver Dis 12:693–712
Storni T, Bachmann MF (2004) Loading of MHC class I and II presentation pathways by exogenous antigens: a quantitative in vivo comparision. J Immunol 17:6129–6135
Strassburg MA, Greenland S, Sorvilla FJ et al (1986) Influenza in the elderly: report of an outbreak and review of vaccine effectiveness reports. Vaccine 4:38–44
Streatfield SJ, Howard JA (2003) Plant-based vaccines. Int J Parasitol 33:479–493
Tremblay M-H, Majeau N, Laliberté Gagné M-E et al (2006) Effect of mutations K97A and E128A on RNA binding and self assembly of papaya mosaic potexvirus coat protein. FEBS J 273:14–25
Turpen TH, Reini SJ, Charoenvit Y et al (1995) Malarial epitopes expressed on the surface of recombinant tobacco mosaic virus. Biotechnology 13:53–57
Vaine M, Lu S, Wang S (2009) Progress on the induction of neutralizing antibodies against HIV type 1 (HIV-1). BioDrugs 23:137–153
Yusibov V, Kumar A, North A et al (1996) Purification, characterization, assembly and crystallization of assembled alfalfa mosaic virus coat protein expressed in Escherichia coli. J Gen Virol 77:567–573
Yusibov V, Mett V, Mett V et al (2005) Peptide-based candidate vaccine against respiratory syncytial virus. Vaccine 23:2261–2265
Zayakina O, Arkhipenko M, Smirnov A et al (2009) Restoration of potato virus X coat protein capacity for assembly with RNA after His-tag removal. Arch Virol 154:337–341
Zeisel MB, Fafi-Kremer S, Fofafa I et al (2007) Neutralizing antibodies in hepatitis C virus infection. World J Gastroenterol 13:4824–4830
Zhao X, Fox JM, Olsen NH et al (1995) In vitro assembly of cowpea chlorotic mottle virus from coat protein expressed in Escherichia coli and in vitro transcribed viral cDNA. Virology 207:486–494

Applications of Plant Viruses in Bionanotechnology

George P. Lomonossoff and David J. Evans

Abstract The capsids of most plant viruses are simple and robust structures consisting of multiple copies of one or a few types of protein subunit arranged with either icosahedral or helical symmetry. In many cases, capsids can be produced in large quantities either by the infection of plants or by the expression of the subunit(s) in a variety of heterologous systems. In view of their relative simplicity, stability and ease of production, plant virus particles or virus-like particles (VLPs) have attracted attention as potential reagents for applications in bionanotechnology. As a result, plant virus particles have been subjected to both genetic and chemical modification, have been used to encapsulate foreign material and have, themselves, been incorporated into supramolecular structures.

Contents

G. P. Lomonossoff (✉) · D. J. Evans
Department of Biological Chemistry, John Innes Centre,
Norwich Research Park, Norwich, NR4 7UH, UK
e-mail: george.lomonossoff@jic.ac.uk

D. J. Evans
e-mail: dave.evans@jic.ac.uk

Current Topics in Microbiology and Immunology (2011) 375: 61–87
DOI: 10.1007/82_2011_184

Published Online: 29 October 2011

1 Introduction

The majority of plant viruses have particles that are nonenveloped and consist of only one or a few types of subunit arranged either with icosahedral or helical symmetry around a single-stranded RNA genome. In many cases plant virus particles can be produced in large quantities in plants and are generally very stable. Virus-like particles (VLPs) also can be produced by expressing the coat protein in a variety of heterologous systems, such as *Escherichia coli*, yeast or insect cells. The advent of techniques for the manipulation of RNA genomes coupled with a detailed knowledge of their virion structure and the use of heterologous expression systems has meant that plant viruses have been at the forefront of the exploitation of virus particles for applications in both bio- and nano-technology. As plant viruses are clearly biological entities which can also be considered as nanoparticles, it is perhaps best to consider plant virus particles as being used in bionanotechnology.

Plant viruses to date have been exploited in three different ways in bionanotechnology: modification of the outer capsid surface (either genetically, chemically or by a combination of both), filling of the inner cavity, and the incorporation of particles into supramolecular structures. This chapter is therefore divided into sections dealing separately with the modification of the outer surface, the incorporation of materials within particles and the incorporation of particles within higher order structures. However, it should be noted that particles with modifications to their outer capsid surface can be used to incorporate foreign material and can, themselves, be incorporated into higher order structures.

2 Modification of the Outer Surface of Plant Virus Particles

The first examples of the modification of the outer surface of virus particles involved the genetic modification of the coat protein (Lomonossoff and Johnson 1996). The original motivation for this work was to modify particles to express antigenic peptides; such modified particles (chimaeras) could potentially serve as novel subunit vaccines. Subsequently, the alternative approach of chemically modifying particles was explored. Several of the amino acids within viral coat protein subunits have side chains which are suitable for chemical modification. Prominent among these are the ε-amino group of lysine, the carboxyl groups of aspartic and glutamic acid, the thiol group of cysteine and the hydroxyl group of tyrosine. When such side chains are exposed on the outer surface of the virus particle, they are addressable by a number of chemical reactants allowing the virus particles to be modified in vitro. This allows the introduction of a greater range of moieties than it is possible to introduce genetically. It is also possible to combine genetic and chemical modification of the virus surface by the introduction or elimination of defined amino acids, thereby modulating the reactivity of the virus particles. It is also possible to genetically insert peptides which catalyse certain reactions, such as the specific deposition of minerals. The capsids of a number of plant viruses with varying morphologies have been modified, both genetically and chemically, on their exterior surfaces. The main prerequisites for genetic modification are that the presence of the foreign sequence does not interfere with the ability of the modified coat protein to assemble into virions or VLPs and that the inserted sequence is, indeed, displayed on the outer surface of assembled particles. For chemical modification, it is important the reaction conditions are not so harsh that they disrupt the virus structure and that some information is available about the numbers and types of addressable groups. For these reasons, attention to date has focussed on those viruses which are known to be robust and for which there is at least some information available about the topology of the coat protein in the assembled virions.

2.1 Icosahedral Viruses

2.1.1 Cowpea Mosaic Virus

Cowpea mosaic virus (CPMV) was the first plant virus to be developed as a system for the display of foreign peptides (Usha et al. 1993; Porta et al. 1994, 1996) and has subsequently been used extensively for chemical modification (Steinmetz et al. 2009a). CPMV is a bipartite RNA virus, with the particles consisting of 60 copies each of a Large (L) and a Small (S) coat protein (CP) arranged with icosahedral symmetry in a particle of approximately 28 nm diameter (Fig. 1). Both coat proteins are produced by proteolytic cleavage of a precursor (VP60) by the virus-encoded 24K proteinase (Franssen et al. 1982; Saunders et al. 2009). The virus particles are attractive candidates for both genetic and chemical modification for a number of reasons: they can be purified in large quantities from infected tissue, they are very robust, surviving at 60°C for at least one hour, across the range of pH 4–10, and in some organic solvent–water mixtures, and their 3-dimensional structure is known to atomic resolution (Lin and Johnson 2003). Furthermore infectious cDNA clones of both RNAs have been available for the past 20 years enabling precise genetic changes to be introduced. The availability of a detailed 3-dimensional structure of the virus particles initially enabled a rational choice to be made regarding potential sites for the insertion of heterologous peptides into the viral coat proteins such that they would be surface-exposed and would not adversely affect particle assembly (Lomonossoff and

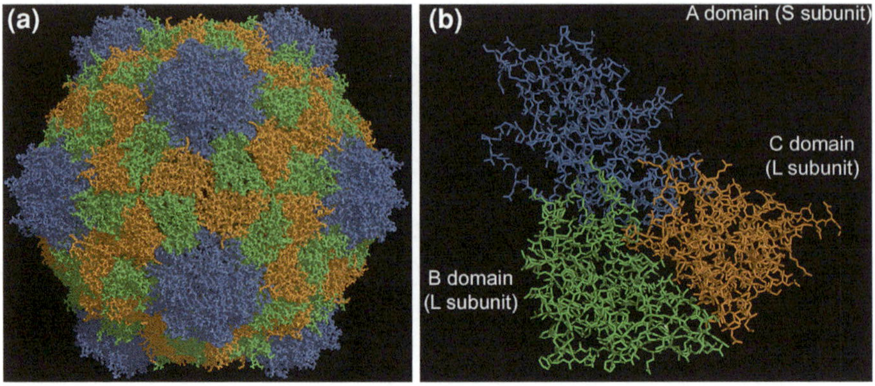

Fig. 1 Structure of **a** the Cowpea mosaic virus (CPMV) capsid and **b** the asymmetric unit. The capsid is comprised of the small(S) and large(L) subunit. (Steinmetz and Evans 2007)

Johnson 1995; Johnson et al. 1997) and has subsequently proved invaluable for making precise chemical modifications (Steinmetz et al. 2009a).

The initial studies on the genetic alteration of the outer surface of CPMV particles were aimed at creating antigenic chimaeras for vaccine purposes. A number of sites were identified as suitable for the insertion of foreign peptides. In most cases, the foreign peptide has been inserted into the most exposed loop of the virus surface, the βB-βC loop of the S protein. However, other sites, such as the βE-αB loop of the L protein and the βC'-βC" loop of the S protein, have also been used successfully (Brennan et al. 1999; Taylor et al. 2000; Chatterji et al. 2002; Porta et al. 2003). Generally, provided the inserted peptide was less than 40 amino acids and had a pI below 9.0 (Porta et al. 2003) the yields of modified particles were similar to those obtained with wild-type CPMV (up to 1mg of particles per gram of infected leaf tissue). In each case, the chimaeric virus particles present 60 copies of the inserted peptide on the virus surface, though preliminary experiments indicate that it will be possible to utilise more than one insertion site simultaneously. Work on the production of chimaeras for vaccine purposes culminated in the demonstration of protective immunity in target animals (Dalsgaard et al. 1997; Langeveld et al. 2001) and the ability to correlate the structure that a peptide adopts with its immunological properties (Lin et al. 1996; Taylor et al. 2000). Work on the immunological properties of antigenic chimaeras is outside the scope of the current chapter and for further information the reader is referred to specialist reviews on the subject (Lomonossoff and Hamilton 1999; Cañizares et al. 2005; Lomonossoff, 2005; Sainsbury et al. 2010).

More recently, the ability to express peptides on the CPMV surface has been exploited to introduce peptides which are capable of, or promote, subsequent chemical modification. Medintz et al. (2005) introduced a hexa-histidine at various points on the virus surface and the modified particles were bound via the introduced histidines to glass slides coated with nickel-nitrilotriacetic acid (Ni–NTA).

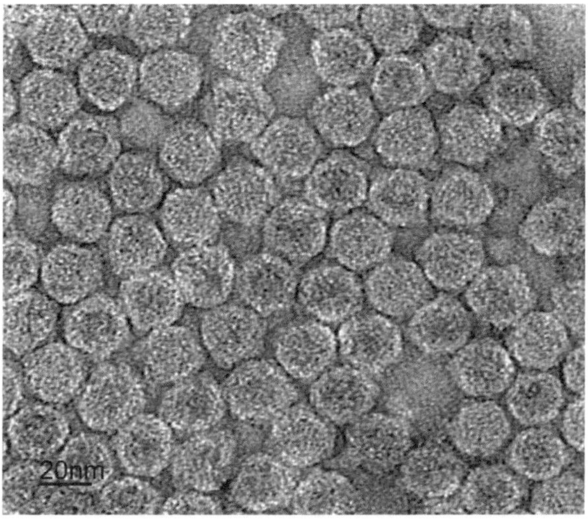

Fig. 2 Transmission electron micrograph of *unstained* FePt-coated CPMV showing monodisperse mineralised particles

The exposed histidine sequences on the upper virus surface were then used to bind a bifunctional Ni–NTA-biotin reagent which allowed the subsequent attachment of quantum dots via the biotin-avidin interaction. In addition, it has been possible to exploit the observation that short peptides can promote specific mineralisation of virus particles. CPMV chimaeras have been produced in which peptides designed to promote the deposition of silica or an iron-platinum (FePt) alloy on the virion surface were expressed in the βB-βC loop of the S protein (Steinmetz et al. 2009b; Shah et al. 2009). In the case of the CPMV-silica chimaera, silication was achieved via a sol–gel process. The silicated CPMV particles could be easily visualised in unstained electron microscopy images. The diameter of silicated particles was estimated to be $\sim$32 nm, consistent with an average silica coating on each particle of 2 nm. In a similar approach, Shah et al. (2009) created a CPMV chimaera that promotes the formation of an FePt alloy on the capsid surface to give monodisperse, hollow, FePt nanoparticles of about 30 nm diameter (Fig. 2). The mineralised particles produced in this way are highly monodisperse and are not readily prepared by other synthetic methods. Further, the process is environmentally benign; as it proceeds at room temperature and pressure, requires no organic solvents, and produces little waste.

Wild-type CPMV particles have five exposed lysines and eight or nine exposed carboxylates (from aspartic and glutamic acid residues) per asymmetric unit, (the asymmetric unit consists of one copy each of the L and S protein) (Fig. 1). This means that each particle should have 300 addressable amine and 480–540 addressable carboxyl groups per virus particle. There are also exposed tyrosines, but no cysteines. All the naturally occurring addressable groups have been exploited to produce chemically modified particles. Probably the most frequently utilised group has been the ε-amino group of surface-exposed lysines. Initial studies revealed that approximately 240 dye molecules could be attached per particle under forcing conditions (Wang et al. 2002a, b) suggesting that four of the five exposed lysines can be modified. The ability to modify lysine residues has since been extensively exploited for a number of purposes. The first example of this was reported by Raja et al. (2003) who derivatised the lysine residues of CPMV with polyethylene glycol (PEG). PEG is known to moderate the immune response to a particle, a phenomenon which may be of considerable use if plant viruses are to be used as novel vaccines. The reaction was shown to give conjugates with altered densities and immunogenicities, consistent with the known chemical and biological properties of PEG. These studies were subsequently refined by Steinmetz and Manchester (2009) who showed that only longer PEG chains effectively shield particles from interacting with cells or tissues. CPMV particles which have been fluorescently labelled at exposed lysines have been successfully used for high-resolution vascular imaging (Lewis et al. 2006). This application makes use of the observations that CPMV particles are nontoxic and rapidly cleared when administered to mammals (Singh et al. 2007) and the propensity of CPMV particles to bind to endothelium surface-expressed vimentin (Koudelka et al. 2009). The exposed lysines have also been extensively used to couple biotin to the virus surface to enable the particles to bind to avidin (Medintz et al. 2005; Steinmetz et al. 2006a) and this introduced binding ability has been exploited for the creation of supramolecular structures (see Sect. 4.1.2). Lysines have also been modified with ferrocenecarboxylate to produce redox-active nanoparticles bearing approximately 240 ferrocene moieties per particle (Steinmetz et al. 2006b). Electrochemical measurements showed that the multiple, redox-active, organometallic centres behave as electronically isolated units. These studies have subsequently been refined and the effect of the length of the linker between the virus surface and the ferrocene on the properties of the resulting particles assessed (Aljabali et al. 2010a) (Fig. 3); the ability to vary this length will be of importance in the design of CPMV-based bioelectronic components and devices. The redox-active particles resulting from these studies may lead to the development of electron-transfer mediators in redox catalysis, amperometric biosensors, and eventually to nanoelectronic devices such as molecular batteries.

A further refinement of the use of addressable lysines has been the combination of their modification with advanced conjugation chemistries such as Cu(I)-catalysed azide-alkyne1,3-dipolar cycloaddition (CuCAAC) reactions ("click" chemistry). Click reactions are a particularly useful strategy for bioconjugation because of their specificity (Strable and Finn, 2009). This approach has been used to attach a variety of molecules, including fluorescent dyes (Meunier et al. 2004; Sen Gupta et al. 2005a), metal complexes (Prasuhn et al. 2007), sugars and polymers (Sen Gupta et al. 2005a, b; Kaltgrad et al. 2007; Astronomo et al. 2010), fullerenes (Steinmetz et al. 2009c) and chemically sensitive RNA fragments (Hong et al. 2009), to the surface of CPMV. The use of this bioorthogonal ligation chemistry with polyvalent display of carbohydrates on the virus surface is an efficient method for the production of anti-glycan polyclonal antibodies with the potential for applications in diagnostics and immunotherapeutics (Kaltgrad et al. 2007).

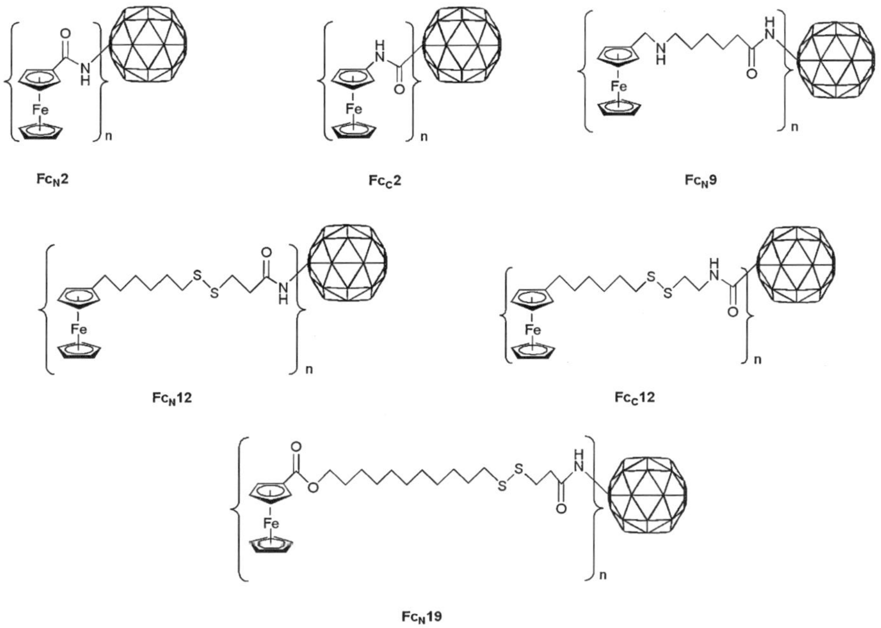

Fig. 3 Redox-active ferrocene-CPMV nanoparticles prepared with various linker lengths and coupling strategies. (Aljabali et al. 2010a)

Though for many purposes modification of the majority of the lysine residues on the CPMV surface is appropriate, others require a greater degree of selectivity. To investigate the reactivity of the various lysines and to examine the possibility of selectively removing some of them, Chatterji et al. (2004) created a series of mutants in which the exposed lysines were substituted with arginines. The results showed that all the lysine residues identified as being exposed are, indeed, addressable and contribute to the overall reactivity of the virus particles. The ability of particles in which all but one of the exposed lysines had been eliminated to allow highly specific attachment of a foreign entity was demonstrated by reacting two mutants, each with a single lysine per asymmetric unit, with monosulfo-NHS-Nanogold followed by cryo-electron microscopy of the resulting conjugates. The electron density of the gold particles corresponded only to the position of single remaining lysine residue which differed in the two mutants (Chatterji et al. 2004).

As an alternative to addressing lysines, Steinmetz et al. (2006c) demonstrated that it is possible to couple the redox-active compound viologen via the surface-exposed carboxyl groups of aspartic and glutamic acids. Though examination of the virus surface indicated that each particle should have 480–540 addressable carboxyl groups only 180 viologen moieties per particle were added. Carboxylates have also been used to introduce ferrocenes (Aljabali et al. 2010a) (Fig. 3). The aromatic side-chain of surface-exposed tyrosine residues has also been investigated as an alternative to the more traditional sites for modification. Using the tripeptide NH2-Gly–Gly-His-COOH (GGH) in the presence of nickel acetate and magnesium monoperoxyphthalate, dityrosine crosslinks were introduced into the virus capsids and the procedure could also be used to make covalent attachments to the virion by trapping with a functionalised disulfide (Meunier et al. 2004). It is also possible to simultaneously address different functionalities of the CPMV surface. Recently, Brunel et al. (2010) have developed a hydrazone ligation strategy to produce targeted CPMV particles which display 133 copies of the VEGFR-1 ligand, 55 copies of a PEGylated peptide, and a total of 188 fluorescent dyes on the virus surface.

In addition to utilising naturally occurring amino acids on the CPMV surface, additional reactive sites can be introduced by mutagenesis. Thiol-addressable CPMV mutants with cysteine residues on the exterior surface were generated by insertion of cysteine residues at specific points on the virus surface (Wang et al. 2002c). The mutant particles reacted with monomaleimido-Nanogold and the derivatised particles imaged by cryo-electron microscopy. The gold particles could clearly be seen at the positions of the inserted cysteine residues, demonstrating that the CPMV particle can function as a convenient and programmable platform for chemical reactions (Wang et al. 2002c) (Fig. 4). Gold nano-particles attached to the surface of cysteine-substituted CPMV particles have been interconnected using molecular wires to create a 3-D conducting network (Blum et al. 2005). The cysteine-substituted mutants have also proved useful for the conjugation of a number of other moieties (Strable and Finn 2009). The presence of genetically introduced cysteine residues has, however, been found to promote aggregation of the virus, making preparations difficult to store and handle. To overcome this problem, a method of chemically introducing thiol groups has been developed by coupling the versatile, thiol-protected, *N*-succinimidyl-S-acetylthiopropionate to surface lysines. On deprotection, the thiol groups are reactive but the particles do not aggregate over several weeks (Steinmetz et al. 2007). A more complete description of the modifications which have been chemically introduced on CPMV particles can be found in Steinmetz and Evans (2007) and Strable and Finn (2009).

All the studies described above have been conducted using particles, either wild-type or genetically modified, produced by the infection of plants. Approximately 90% of the particles produced in this way contain either RNA-1 or RNA-2 and the preparations retain their ability to infect plants and spread in the environment. In addition, while CPMV RNAs have not been shown to be capable of replication in mammalian cells, uptake of particles does occur both in vitro and in vivo (Steinmetz et al. 2009a), raising biosafety concerns. To address these issues, there have been attempts to inactivate or eliminate the viral RNAs within CPMV preparations either by irradiation with ultraviolet light (Langeveld et al. 2001; Rae et al. 2008) or chemically (Ochoa et al. 2006; Phelps et al. 2007). However, these processes have to be carefully monitored as they risk altering the structural properties of the particles and require that the virus retains its functionality to infect and spread in plants. The recent observation that co-inoculation of plants with the VP60 precursor to the L and S coat protein and the 24K viral proteinase

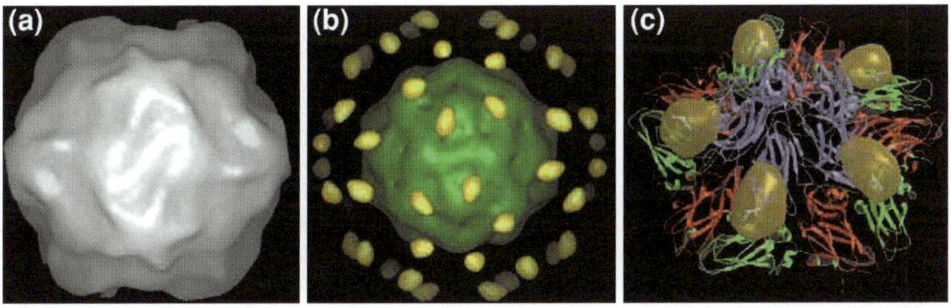

Fig. 4 Cryoelectron microscopy of derivatised CPMV-Cys mutant. **a** Three-dimensional reconstruction of CPMV particles with 1.4 nm nanogold clusters. **b** Difference electron density map showing the nucleic acid (*green*) and the gold particles. **c** A pentameric section of the difference electron density map around the five-fold symmetry axis superimposed on the atomic model of CPMV showing that the gold is attached at the site of the Cys mutation. (Wang et al. 2002c, with permission, Copyright 2002 Wiley–VCH Verlag GmbH & Co. KGaA.)

results in the production of empty (RNA-free) CPMV capsids (Saunders et al. 2009) suggests a way around this problem. Thus many future studies involving the modification of the outer surface may well be conducted using particles produced in this manner rather than by infection. Furthermore, the empty particles produced in this manner could potentially be loaded with foreign "cargo" (see Sect. 3.1.5).

2.1.2 Cowpea Chlorotic Mottle Virus

Cowpea chlorotic mottle virus (CCMV) is a tripartite virus which has a capsid that consists of 180 identical coat protein subunits which form a spherical particle 28 nm in diameter. The CCMV capsid displays addressable lysines and carboxylates derived from aspartic and glutamic acid. Amine and carboxy-selective chemistry has been used to selectively attach fluorescent dyes to the virus surface, with approximately 540 lysine residues and 560 carboxylates being addressable (Gillitzer et al. 2002). It also proved possible to genetically introduce two solvent-exposed cysteines per CP into the virus capsid. Probing the resultant particles with thiol-selective dyes showed that approximately a third of the introduced thiols could be addressed. Subsequently a large diversity of ligands including fluorescent dyes, an organometallic photosensitiser, biotin, small peptides and even intact IgG antibodies can be effectively chemically linked to the exterior surface of CCMV, clearly illustrating that chemical modification is a generic approach to surface modification of the virus capsid (Gillitzer et al. 2002; Suci et al. 2007a, b). Further details of the chemical attachment of molecules to the surface of CCMV can be found in Steinmetz and Evans (2007) and Young et al. (2008).

A particular advantage of CCMV is the ability of the coat proteins subunits, isolated either from virus particles or through the expression of the CCMV coat protein in heterologous expression systems, to assemble into VLPs in vitro. Using

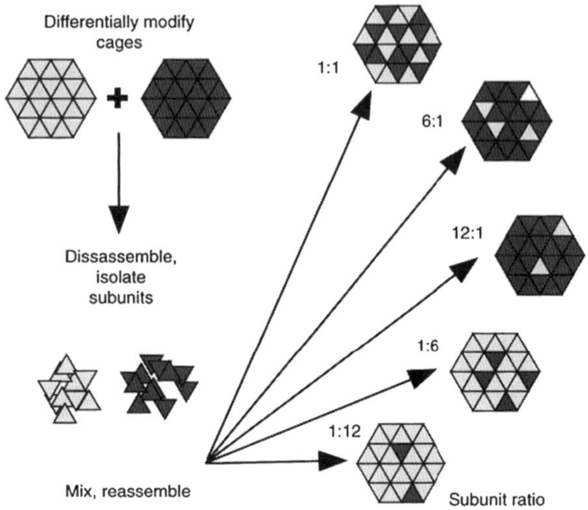

Fig. 5 Schematic for the assembly of multifunctionalised Cowpea chlorotic mottle virus particles. (Gillitzer et al. 2006, with permission, Copyright 2006 Wiley–VCH Verlag GmbH & Co. KGaA.)

this ability, a method has been developed that allows hybrid CCMV particles containing differentially modified subunits to be produced through mixed self-assembly (Gillitzer et al. 2006). To achieve this, CCMV particles were independently decorated with two different types of ligands to generate two populations of labelled virions. The two types of particles were disassembled in vitro and the resulting subunits separately purified. Reassembly was performed using varying ratios of the two types of subunits, thereby controlling the stoichiometries of the two ligands on the final assembled virions (Fig. 5).

2.1.3 Turnip Yellow Mosaic Virus

Turnip yellow mosaic virus (TYMV) is the type member of the genus *Tymovirus*. Virions consist of 180 identical coat protein subunits per particle; however only 60 lysines per virus particle (one per 3 subunits) could be readily addressed using standard reagents (Barnhill et al. 2007). In addition, 90–120 carboxyl groups, located in the most exposed region of the coat protein, could also be modified.

2.2 Rod-Shaped Viruses

2.2.1 Tobacco Mosaic Virus

Particles of TMV consist of a single molecule of genomic RNA encapsidated by 2130 copies of the 17.5 kDa coat protein arranged with helical symmetry to form rigid particles 300nm in length. The subunits are largely α-helical, with the N- and C-termini being exposed on the outer virus surface. TMV is probably the highest yielding plant virus and thus its particles have attracted considerable interest for both peptide presentation via genetic modification and direct chemical modification. Most attempts to express foreign peptides via genetic fusion have focussed on the C-terminus of the coat protein in view of its exposed location. However, although TMV particles contain a large number of subunits, making the system potentially very attractive for peptide expression, it also creates a problem in that the subunits are very tightly packed, allowing little space on the virus surface for the expression of foreign sequences. A number of strategies have been developed to address this problem. Hamamoto et al. (1993) developed a TMV vector which permitted the synthesis of both native and C-terminally modified versions of the coat protein from the same viral RNA. This was achieved by engineering a leaky termination codon at the C-terminus of the coat protein gene. This system produced particles in plants in which up to 5% of the coat protein subunits were modified at their C-termini and has been used to express epitopes from several animal pathogens (Sugiyama et al. 1995; Turpen et al. 1995). Subsequently, by modifying the site of peptide insertion, TMV-based peptide presentation systems have been developed in which all the coat protein subunits could be modified to express foreign peptides without abolishing virus viability (Turpen et al. 1995; Fitchen et al. 1995). Using a similar approach, Koo et al. (1999) showed that mice immunised with a TMV chimaera expressing a peptide from the spike protein of the coronavirus, Murine hepatitis virus (MHV), were protected from subsequent challenge with the virus.

A problem with using direct fusions of peptides at or near the C-terminus of the TMV coat protein is that the size of inserts which can be tolerated seems to be quite small, the largest reported insert at this site being 23 amino acids (Bendahmane et al. 1999). However, this problem can be alleviated by modifying the C-terminus of the coat protein. Werner et al. (2006) found that a functional fragment of protein A, 133 amino acids in length, could be displayed on the surface of a close relative of TMV, the tobamovirus Turnip vein clearing virus (TVCV), if the sequence was fused to the C-terminus of the coat protein via a 15-amino acid linker. The protein A fragment was able to bind IgG suggesting that the modified virus particles could be used to purify antibodies. Given the success of TMV-based systems for peptide presentation, there has been considerable interest in the commercial development of the technology. For a description of such developments, the reader is referred to recent reviews by McCormick and Palmer (2008) and Smith et al. (2009).

The outer surface of wild-type TMV particles is somewhat devoid of chemically reactive amino acids such as cysteine and lysine. However, Schlick et al. (2005) reported the derivatisation of exterior-exposed tyrosine residues with diazonium salts, resulting in an acetophenone-functionalised conjugate that could react with a wide range of other molecules including PEG monomethyl ethers. To overcome the lack of reactive amino acid chains on the virus surface, several mutants displaying reactive cysteine or lysine residues on the solvent-exposed exterior of the virus have been made, allowing decoration via thiol- or amine-selective chemistry (Demir and Stowell 2002; Yi et al. 2005, 2007). However, in many cases the presence of these added residues adversely affected virus yield.

To counteract this, Smith et al. (2006) screened a random collection of TMV mutants which had an additional four amino acids, including a single lysine, inserted near the N-terminus of the coat protein. By selecting those mutants which grew well, the authors were able to identify a particular mutant which could be used for the chemical coupling of a variety of epitopes (McCormick and Palmer 2008).

The external surface of TMV has also been used as a template for deposition reactions including the cocrystallisation of CdS and PbS, oxidative hydrolysis (iron oxides), and sol–gel condensation (SiO_2) (Shenton et al. 1999; Fowler et al. 2001; Royston et al. 2009) and for the creation of metal nanoparticles and nanotubes (Dujardin et al. 2003; Knez et al. 2004a; Bromley et al. 2008). Such tubes can be grown from deposited clusters of palladium, platinum and gold on the exterior TMV surface and result in a metallic coat on the virion that serves as a basal layer for electroless deposition of other metals including nickel and cobalt. Royston et al. (2008) reported the deposition of nickel and cobalt on the exterior surface of the virus to create metallic coatings up to 40 nm in thickness. When the nickel-coated virions were incorporated into a nickel–zinc battery system, the electrode capacity of the battery more than doubled. For a more detailed review the reader is referred to Evans (2007).

2.2.2 Potato Virus X

Potato virus X (PVX) has filamentous particles consisting of approximately 1260 coat protein subunits encapsidating a single RNA molecule. Though an atomic resolution structure of the coat protein subunits is not available, the overall architecture of the viral particles is known (Kendall et al. 2008). It has proved possible to genetically fuse peptides to the surface- exposed N-terminus of either a proportion of or all of the subunits. To achieve partial modification, the sequence of the FMDV 2A catalytic peptide was inserted between the peptide and the N-terminus of the coat protein (Santa Cruz et al. 1996) such that both wild-type and N-terminally modified subunits can be produced from the same construct. This approach has the advantage that it potentially permits the expression of longer peptides, including whole proteins, than would be the case if all the subunits are modified. Using the ability of the 2A cleavage strategy to permit the fusion of whole proteins, Smolenska et al. (1998) expressed a single-chain antibody on the particle surface and showed that it retained its binding specificity. Carette et al. (2007) subsequently expressed the enzyme lipase B from *Candida antarctica* on the surface of the virus. These authors showed that the virus-anchored lipase molecules were catalytically active and suggested that it could act as an anchored biocatalyst.

Using an alternative approach of modifying all the subunits, Marusic et al. (2001) expressed a highly conserved hexapeptide epitope from gp41 of HIV-1 on PVX particles. Mice immunised with the chimaeric particles produced high levels of HIV-1-specific IgG and IgA. To determine the rules which govern the peptides that could be successfully propagated as fusions to the N-terminus of the PVX coat protein, Lico et al. (2006) undertook a detailed analysis of a large number of PVX

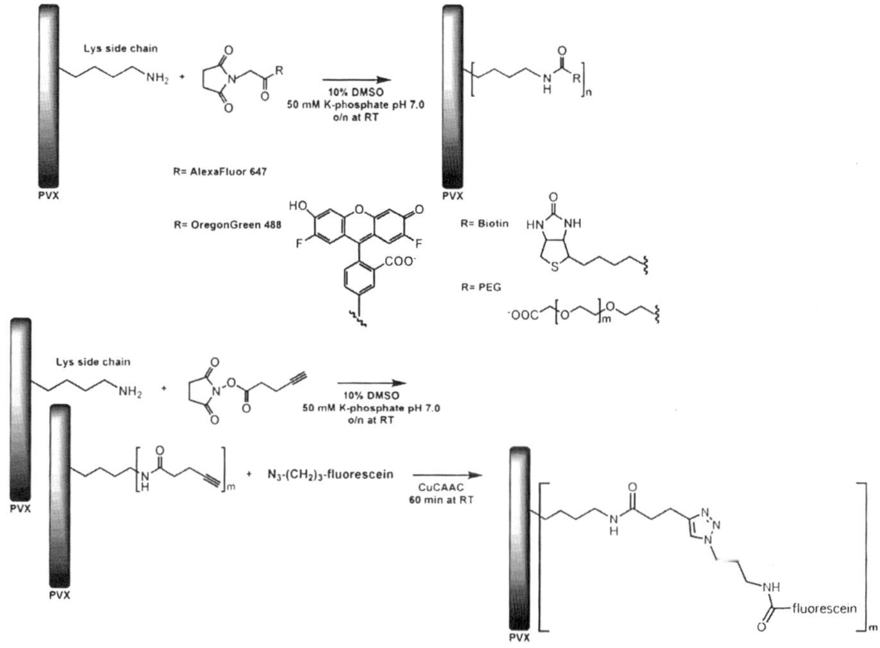

Fig. 6 Chemical labelling of Potato virus X with fluorescent dyes and PEG chains using NHS ester-based chemistry or click reactions. (Steinmetz et al. 2010, with permission, Copyright 2010 American Chemical Society)

chimaeras with N-terminal extensions of different lengths and amino acid composition. The results showed that the isoelectric points and tryptophan content of the inserted peptides had a profound influence on the growth of the chimaeras; this knowledge has subsequently been exploited to produce chimaeras in a predictable manner (e.g., Lico et al. 2009). However, there do appear to be occasions when modification of all the subunits is less successful than the partial modification strategy even for short peptides (Uhde-Holzem et al. 2010).

To examine the possibility of chemically modifying PVX, Steinmetz et al. (2010) conducted a detailed study of the reactivity of functional groups present on the surface of the particle. Each of the 1260 PVX CP subunits contains 11 lysine residues, 10 aspartic acid residues and 10 glutamic acid residues, all of which could potentially be modified if they were surface-exposed. In addition, the subunits of some strains of PVX are glycosylated, and the glycan moieties are also potential targets for chemical modification. Preliminary data indicated that none of the carboxylates were addressable under the conditions used and that the coat protein of the particular strain of PVX used for the experiments was not glycosylated. By contrast, lysine residues could be modified with an average of just over one lysine per subunit being modified. The utility of lysine residues for the introduction of PEG molecules of differing sizes and participation in click chemistry was subsequently investigated (Fig. 6).

3 Utilising the Interior of Plant Virus Particles

The interiors of plant virus capsids potentially provide a nanosize environment for the packaging of foreign materials. There are essentially two approaches that can be taken to encapsulate foreign molecules within the capsid. In the first approach, the foreign molecules are incorporated into the particles during the capsid assembly process. In the second, the foreign molecules are introduced into pre-assembled particles. Most research has concentrated on utilising the enclosed space of a variety of icosahedral viruses; however the internal channel of TMV, which is open at both ends, has also be used for some specific purposes. In yet another type of application, the ability of the virus coat protein to package specific RNA molecules has been exploited.

3.1 Icosahedral Viruses

3.1.1 Cowpea Chlorotic Mottle Virus

The plant virus which has been most extensively used for interior modification is CCMV. Empty (RNA-free) particles of CCMV can be produced in vitro through the assembly of coat protein subunits. CCMV particles have a particular advantage for the encapsulation of foreign molecules as they undergo a pH- and cation-dependent structural transition that can be used to control the loading and release of such material. At pHs above 6.5 and in the absence of divalent cations, the CCMV capsid undergoes a reversible swelling which increases the diameter of the particles by about 10% and leads to the formation of 60 pores (Speir et al. 1995). The interior surface of wild-type particles carries a high positive charge density due to the presence of nine basic residues (arginine and lysine) in the amino-terminal region of each subunit. These positively charged residues normally interact with the negatively charged viral RNA. The positively charged interior surface and the availability of pores have been used to promote mineralisation within the preformed capsid to produce defined inorganic nanoparticles of anionic polyoxometallate salts (Douglas and Young 1998) or titania particles (Klem et al. 2008). The resulting nanoparticles were constrained in both size and shape by the interior dimensions of the CCMV virion. By a photoinitiated stepwise reaction, encapsulated precursor iron complexes within CCMV can be converted to monodisperse Prussian blue nanoparticles (de la Escosura et al. 2008). Further, these CCMV-internalised Prussian blue nanoparticles can be organised on mica and other substrates, assisted by the propensity of CCMV to self-assemble into hexagonal patterns, opening the way for exploitation of their magnetic and optical properties. Using heterologous expression, it has proved possible to produce CCMV particles with an altered interior charge. When the interior was made acidic, it proved possible to catalyse the formation of cationic transition metal oxides inside the particles when they were incubated with the appropriate cations

(Douglas et al. 2002). In an alternative approach, the endogenous calcium-binding sites present at the interface of the coat protein subunits in the assembled CCMV capsid were used to bind up to 180 gadolinium (Gd^{3+}) ions (Liepold et al. 2007). The binding of Gd^{3+} ions to the CCMV capsid results in particles which are paramagnetic and which have potential as MRI contrast agents in vivo.

The CCMV capsid has also shown to be capable of packaging enzymes and polymers. Comellas-Aragonès et al. (2007) demonstrated that one horseradish peroxidase (HRP) molecule, which is about 40 kDa, could be packaged within the CCMV capsid during an in vitro assembly reaction. The packaged HRP remained enzymatically active and substrate accessibility could be modulated by the pH-dependent swelling that CCMV capsids undergo. This result suggests that it may be possible to design nanoreactors based on controlled substrate access to encapsulated enzymes. Subsequently, it was shown (Comellas-Aragonès et al. 2009) that the synthetic polymer, polystyrene sulfonate, can be incorporated within capsids which were modified with PEG on their external surface. Most recently, a strategy has been developed for the self-assembly and loading of CCMV capsids using DNA amphiphiles. The amphiphiles aggregate into micelles with a hydrophobic core and an anionic DNA corona; these negatively charged particles induce capsid formation. By either hybridizing small molecules onto the micelles or incorporating them into the core, co-packaging of various small compounds can be achieved (Kwak et al. 2010). It is anticipated that after further development such nanocarriers may act as high-impact drug delivery systems.

Another potentially useful feature of CCMV is the ability of its coat protein to assemble into structures distinct from the normal virion. For example, it is possible to produce particles containing 60 or 120 subunits, as opposed to 180, by making deletions in the N-terminus of the coat protein (Tang et al. 2006). In an even more dramatic illustration, DNA has also been shown to nucleate the assembly of tube-like structures composed of CCMV coat protein dimers (Mukherjee et al. 2006).

3.1.2 Brome Mosaic Virus

Brome mosaic virus (BMV) is in the same genus as CCMV and BMV particles have similar structure and properties. BMV coat protein has been shown to be able to assemble around preformed gold nanoparticles provided there is a citrate layer between the gold and the protein surface (Dragnea et al. 2003). The encapsulation efficiency of the gold nanoparticles could be improved by coating them with carboxyl terminated polyethylene glycol (Chen et al. 2006); the size of the resultant capsid was found to be influenced by the diameter of the nanogold particle (Sun et al. 2007). Quantum dots have also been encapsulated within BMV particles using in vitro assembly reactions (Dixit et al. 2006). Fluorescence measurements of the material suggested that viral particles containing encapsulated quantum dots could be developed into high-performance luminescent probes. Using a similar approach, Huang et al. (2007) assembled BMV capsids around iron oxide nanotemplates. When the iron oxide core was larger than the inner cavity of native BMV, capsids larger than native BMV particles were obtained. The particles containing the iron oxide were superparamagnetic, suggesting that they could have applications in magnetic imaging and biosensing.

3.1.3 Red Clover Necrotic Mosaic Virus

Red clover necrotic mosaic virus (RCNMV) is a bipartite RNA virus which has particles consisting of 180 identical subunits arranged with icosahedral symmetry and with a diameter of approximately 36 nm. Virus particles are stabilised by an internal protein-RNA cage and their assembly is initiated with the recognition of an origin of assembly site on the viral RNA by the CP (Sit et al. 1998). By attaching an artificial origin of assembly sequence to a gold nanoparticle, it proved possible to achieve the in vitro encapsulation of the gold particle by the viral CP (Loo et al. 2006, 2007). Using this approach, it proved possible to encapsulate a range of gold core sizes. As suggested for BMV, the resultant material could be used for biosensing purposes and RCNMV has been proposed as a targeted particle for cancer treatment (Franzen and Lommel 2009).

3.1.4 Hibiscus Chlorotic Ringspot Virus

Hibiscus chlorotic ringspot virus (HCRSV) is a member of the genus *Carmovirus*. Virions have 28 nm particles which consist of 180 identical coat protein subunits; empty particles can be produced by the disassembly/reassembly of virions. Anionic polymers, such as polystyrenesulfonic acid and polyacrylic acid, but not neutrally charged dextran molecules, could be successfully loaded into these empty particles (Ren et al. 2006). Ren et al. (2007) made use of this phenomenon to co-encapsulate the anti-cancer drug doxorubicin with polystyrenesulfonic acid into HCRSV particles. To target the particles to cancerous cells, folic acid was conjugated to lysine residues on the outer virus surface. The resultant particles improved the uptake and cytotoxicity of doxorubicin to ovarian cancer cells suggesting that modified plant virus capsids may provide the basis for targeted drug delivery in cancer chemotherapy.

3.1.5 Cowpea Mosaic Virus

Until recently the interior cavity of CPMV particles has not been amenable to loading with foreign molecules as no in vitro assembly system has been available and the production of virions via the infection of plants results in the majority of particles containing the viral RNA. An initial attempt to produce RNA-free, loadable particles involved treating wild-type CPMV at high pH to eliminate the encapsidated virion RNAs (Ochoa et al. 2006). The potential utility of the resultant RNA-free particles was demonstrated by showing that cysteine residues on the inner capsid surface, which are normally occluded by the viral RNA, could now be labelled with a reporter dye. The alternative approach of producing empty particles

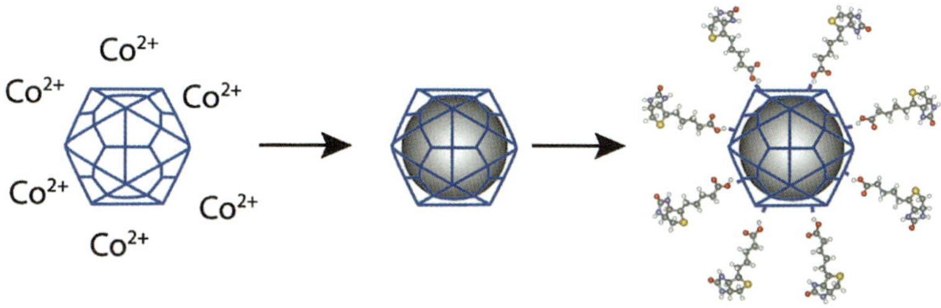

Fig. 7 Schematic of the loading of an empty virus-like particle of CPMV and subsequent chemical decoration of the outer surface

within plants by the co-expression of VP60 and the 24K proteinase (Saunders et al. 2009) has recently been shown to be able to produce particles which are capable of being loaded with cobalt or iron oxide (Aljabali et al. 2010b). The presence of the metal within the particles allows them to be visualised by electron microscopy in the absence of negative stain. Importantly, the external surface amino acids of the loaded-virus capsid remain amenable to chemical modification (Fig. 7). The ability to both encapsulate materials, such as drugs and nanoparticles, within the capsid and to chemically modify the outer surface, opens up routes for the further development of these systems for the targeted delivery of therapeutic agents.

The ability of CPMV to encapsidate different lengths of RNA has been exploited to render heterologous RNA highly resistant to degradation. It had previously been observed that RNA-2-containing particles can accommodate RNA-2 molecules with additional lengths of heterologous sequence. These observations have been used to design modified versions of RNA-2 harbouring pathogen-specific sequences that can act as positive controls in highly sensitive real-time PCR-based diagnostic reactions (King et al. 2007).

3.2 Rod-Shaped Viruses

3.2.1 Tobacco Mosaic Virus

TMV particles are hollow cylinders with an internal diameter of 4 nm. The interior channel is lined with aspartic and glutamic acid residues and these have been labelled with a variety of small molecules, such as biotin, using carbodiimide coupling reactions (Schlick et al. 2005). Nanowires consisting of bimetallic alloys of CoPt, $CoPt_3$ and $FePt_3$ with lengths up to 100 nm and 4 nm diameter have been synthesised within the TMV capsid channel (Tsukamoto et al. 2007) and the formation of small isolated nanoparticles of silver and nickel within the channel has also been reported (Dujardin et al. 2003). Copper nanowire of 3 nm diameter and up to 150 nm length can be formed within the internal channel by electroless deposition (Balci et al. 2006). Further, judicious choice of reaction conditions enables the channel to be filled with rod-shaped nickel or cobalt nanoparticles or

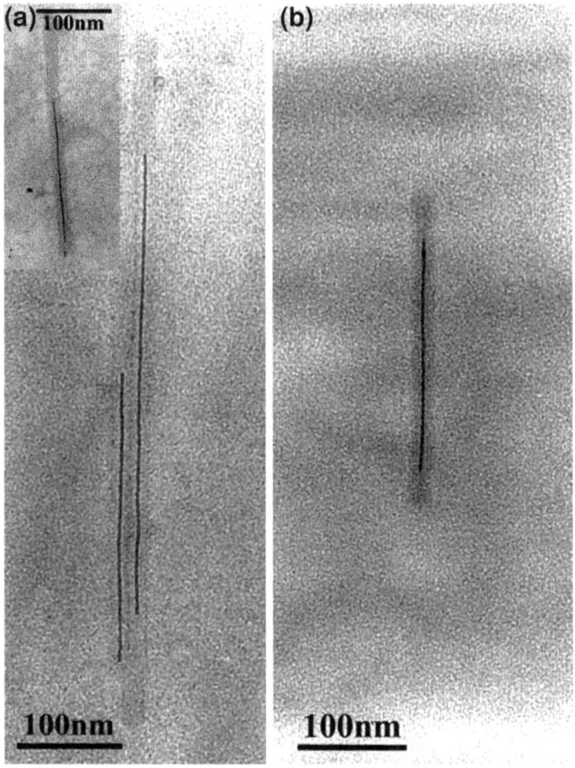

Fig. 8 **a** Transmission electron micrograph of Tobacco mosaic virus showing two adjacent virion aggregates filled with nickel wires. Inset: a single virion filled with a 200 nm long wire. **b** TMV containing a 200 nm long cobalt wire (Knez et al. 2003, with permission, Copyright 2003 American Chemical Society)

with nickel or cobalt wires of 3 nm width and lengths of several hundred nanometres (Knez et al. 2003) (Fig. 8). Lengths greater than the virion length of 300 nm can be achieved, as the virion has a tendency to aggregate end-to-end, often reaching several micrometres.

The encapsidation of TMV RNA by its coat protein is known to proceed from a defined sequence on the viral RNA, the origin of assembly. It has been known for some time that attachment of this sequence to a heterologous RNA will promote encapsidation of the foreign RNA by the coat protein (Gallie et al. 1987). Smith et al. (2007) exploited this phenomenon to deliver RNA encoding the non-structural proteins from Semliki forest virus into mammalian cells. They showed that the encapsidated RNA was uncoated, translated within the cells and stimulated an immune response in mice.

4 Creating Supramolecular Structures

A major aim of nanotechnology is to incorporate nano-sized components into small-scale devices. In the case of virus-based bionanotechnology, this involves the incorporation of modified viruses or virus-like particles into supramolecular structures, often by binding the particle to surfaces. An early example of the incorporation of a plant virus into a supramolecular structure was reported by Lvov et al. (1994) who incorporated the icosahedral particles of Carnation mottle virus (CarMV), the type member of the genus *Carmovirus*, into an alternating multi-layered thin film. With the advent of methods for the genetic and chemical modification of particles, the range of structures that can be built up has steadily become more sophisticated.

4.1 Icosahedral Viruses

4.1.1 Cowpea Chlorotic Mottle Virus

CCMV particles have been immobilised on to surfaces with a view to constructing arrays. Immobilisation was achieved by either adsorption of cysteine-containing CCMV particles on to a gold surface (Klem et al. 2003) or via electrostatic interactions of the negatively charged capsids on to positively charged surfaces (Suci et al. 2005, 2006). Furthermore, multilayers consisting of CCMV particles immobilised on a solid support can be constructed using electrostatic interactions or the biotin–streptavidin interaction. The ability to construct thin films of CCMV could be coupled with the multivalent display of various molecules on the capsid surface and with the ability of the virus to encapsulate and release materials from the capsids in a pH-dependent manner. This could potentially lead to the development of semi-permeable functionalised membranes or controlled release coatings.

4.1.2 Cowpea Mosaic Virus

The first approach to the incorporation of modified CPMV particles into a larger structure involved immobilising particles expressing histidine residues (Medintz et al. 2005). In addition to producing continuous layers, single virus particle arrays have also been constructed (Cheung et al. 2003, 2006, 2010; Smith et al. 2003). Techniques such as microcontact patterning or scanning- and dip-pen nanolithography have been used to generate patterns of functional groups on solid supports that can subsequently be used for binding of the viral particles (Smith et al. 2003).

CPMV can be crystallised leading to the formation of 3-D arrays with ordering to near atomic precision. These crystals contain large solvent channels which can potentially be exploited to allow the diffusion of materials into the crystal interior. The channels have been utilised for confined and regular growth of metals such as palladium and platinum (Falkner et al. 2005). After stabilising the crystals by glutaraldehyde cross-linking, they were exposed to solutions that led to the

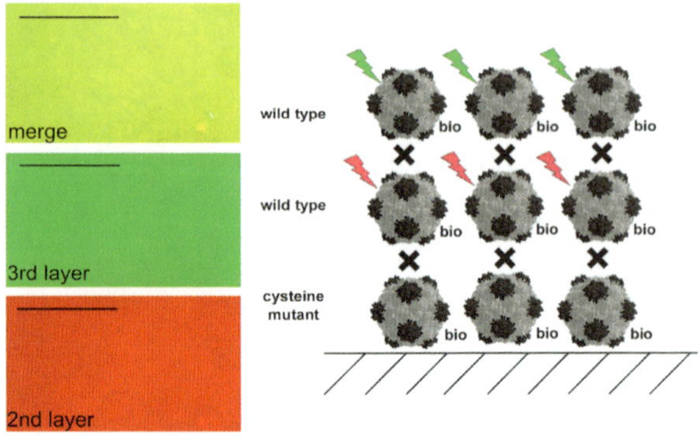

Fig. 9 Triple layer of CPMV particles on a gold surface. Fluorescence microscopy images (*left*) and diagrammatic representation of layer structures (*right*). The *red* and *green* flags show AlexaFluor dyes AF488 and AF568, respectively, and the black cross depicts streptavidin (Steinmetz et al. 2006a, with permission, Copyright 2006 American Chemical Society)

formation of Pd and Pt within the crystal. Transmission electron microscopy studies confirmed that the formation of the metals was confined to the channels.

An alternative method of creating 3-D structures is to build up successive 2-D structures using a layer-by-layer (LbL) approach. To test the feasibility of this with CPMV, particles were covalently labelled with two different ligands: biotin to allow self-assembly via interaction with streptavidin, and fluorescent labels to enable the particles to be imaged (Steinmetz et al. 2006a). Attachment of the different functionalities was achieved via the modification of lysine side chains. The immobilisation of the CPMV particles on a solid support was achieved using either direct binding of cysteine-added mutants to a gold surface or indirectly by binding biotinylated particles mediated via streptavidin. To follow the LbL assembly, biotinylated particles were labelled with different dye molecules. Fluorescence microscopy imaging of the CPMV arrays was consistent with successful binding of successive layers of virus particles (Fig. 9). To study the properties of the multilayers, CPMV LbL assembly was followed in real time and in situ using quartz crystal microbalance with dissipation monitoring (QCMD; Steinmetz et al. 2008a). In particular, the effect of modifying the density of biotin molecules attached via longer and shorter linkers to the virus surface was investigated. These studies revealed that a more regular and densely packed array was produced when CPMV particles displaying a high number of biotin labels attached via the longer linker were assembled. It has also proved possible to build up multilayers of CPMV particles utilising electrostatic interactions (Steinmetz et al. 2008b). Ultrathin films of alternating layers of anionic and cationic polyelectrolytes were built up from a solid surface, by a LbL approach, finishing with a cationic layer. Negatively charged sphere-like CPMV particles were then immobilised on the polyelectrolyte surface. Additional layers of polyelectrolyte-CPMV-polyelectrolyte can readily be introduced (Fig. 10). The incorporation of functionalised virus nanoparticles and polyelectrolytes should lead to new routes for the construction of nano-structured tunable films.

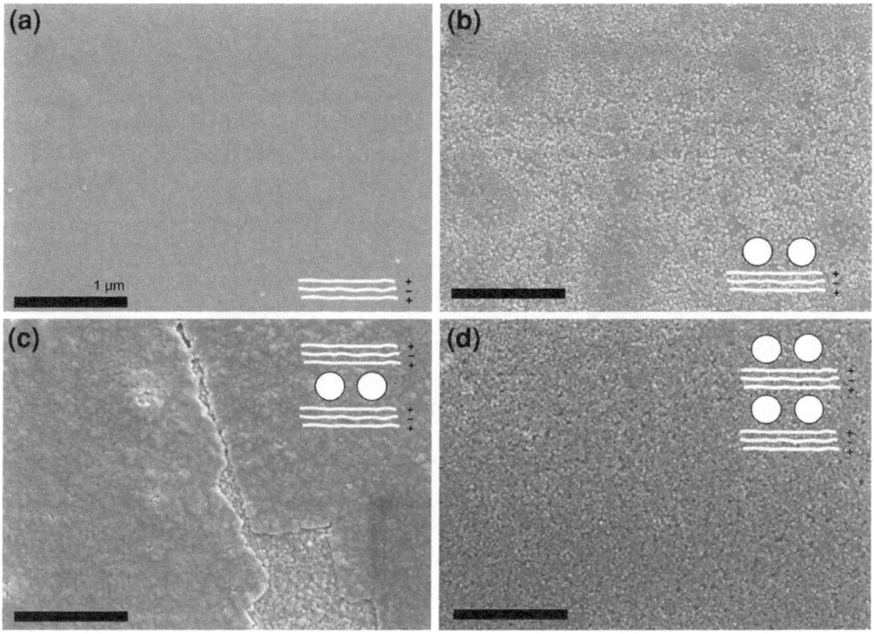

Fig. 10 Scanning electron micrographs showing the sequential build-up of polyelectrolytes and CPMV particles. **a** Precursor polyelectrolyte thin film. **b** Initial polyelectrolyte layer coated with a layer of CPMV particles. **c** Coating of polyelectrolyte-CPMV layer with further polyelectrolyte layers. **d** Subsequent addition of another layer of CPMV particles (Evans 2007, Copyright 2007 with permission from Elsevier)

4.2 Rod-Shaped Viruses

4.2.1 Tobacco Mosaic Virus

The adsorption properties of TMV on various surfaces such as gold, mica, glass and silicon wafers have been investigated (Knez et al. 2004b) and a technique for rapid and large scale assembly of thin film coatings and ordered fibres consisting of aligned TMV particles has also been reported (Kuncicky et al. 2006). Yi et al. (2005, 2007) partially disassembled the coat protein from TMV particles to expose the RNA at the 5′ end of the rods. Oriented assembly of TMV on solid supports was then achieved in a controlled manner via nucleic acid hybridisation using complementary oligonucleotides and the immobilisation of fluorescently labelled TMV onto electrodes was also demonstrated. Furthermore, by using differentially labelled TMV particles and a micropatterned substrate, a patterned TMV micro-array has been constructed. Attempts to incorporate TMV particles into multi-layers using electrostatic interactions revealed that, unlike the spherical CPMV particles, the rods floated on top of the structures (Steinmetz et al. 2008b).

5 Conclusions

The past two decades have seen tremendous advances in the manipulation of plant virus particles, both genetically and chemically, and the investigations into the potential uses to which such modified particles can be put. However, at present all studies have been conducted at an academic level. Thus, a major challenge in the future will be the deployment of the technical advances in both the biotechnology and nanotechnology fields. For example, although it has been demonstrated that chimaeric plant virus particles can stimulate protective immunity in experimental animals, this technology has not been approved for use outside the laboratory. The same is true for the potential imaging agents based on the incorporation of foreign materials within particles. In a similar way, a major step in the adoption of plant virus particles in nanotechnology will require the demonstration that some type of device with unusual or highly desirable properties can be produced cost-effectively.

References

Aljabali AAA, Barclay JE, Butt JN et al (2010a) Redox-active ferrocene-modified cowpea mosaic virus nanoparticles. Dalton Trans. 39:7569–7574

Aljabali AAA, Sainsbury F, Lomonossoff GP et al (2010b) Cowpea mosaic virus unmodified virus-like particles can be loaded with metal and metal oxide. Small 6:818–821

Astronomo RD, Kaltgrad E, Udit AK et al (2010) Defining criteria for oligomannose immunogens for HIV using icosahedral virus capsid scaffolds. Chem Biol 17:357–370

Balci S, Bittner AM, Hahn K et al (2006) Copper nanowires within the central channel of tobacco mosaic virus. Electrochim Acta 51:6251–6357

Barnhill NH, Reuther R, Ferguson PL et al (2007) Turnip yellow mosaic virus as a chemoaddressable bionanoparticle. Bioconjugate Chem 18:852–859

Bendahmane M, Koo M, Karrer E et al (1999) Display of epitopes on the surface of tobacco mosaic virus: impact of charge and isoelectric point of the epitope on virus-host interactions. J Mol Biol 290:9–20

Blum AS, Soto CM, Wilson CD et al (2005) An engineered virus as a scaffold for three-dimensional self-assembly on the nanoscale. Small 1:702–706

Brennan FR, Jones TD, Gilleland LB et al (1999) Pseudomonas aeruginosa outer-membrane protein F epitopes are highly immunogenic in mice when expressed on a plant virus. Microbiology 145:211–220

Bromley KM, Patil AJ, Perriman AW et al (2008) Preparation of high quality nanowires by tobacco mosaic virus templating of gold nanoparticles. J Mater Chem 18:4796–4801

Brunel FM, Lewis JD, Destito G et al (2010) Hydrazone ligation strategy to assemble multifunctional viral nanoparticles for cell imaging and tumor targeting. Nano Lett 10: 1093–1097

Cañizares MC, Lomonossoff GP, Nicholson L (2005) Development of cowpea mosaic virus-based vectors for the production of vaccines in plants. Expert Rev Vaccines 4:687–697

Carette N, Engelkamp H, Akpa E et al (2007) A virus-based catalyst. Nat Nanotechnol 2:226–229

Chatterji A, Burns LL, Taylor SS et al (2002) Cowpea mosaic virus: from the presentation of antigenic peptides to the display of active biomaterials. Intervirology 45:362–370

Chatterji A, Ochoa W, Paine M et al (2004) New addresses on an addressable virus nanoblock: uniquely reactive lys residues on cowpea mosaic virus. Chem. Biol. 11: 855–863

Chen C, Daniel MC, Quinkert ZT et al (2006) Nanoparticle-templated assembly of viral protein cages. Nano Lett 6: 611–615

Cheung L, Camarero JA, Woods BW et al (2003) Fabrication of assembled virus nanostructures on templates of chemoselective linkers formed by scanning probe nanolithography. J Am Chem Soc 125:6848–6849

Cheung CL, Chung SW, Chatterji A et al (2006) Physical controls on directed virus assembly at nanoscale chemical templates. J Am Chem Soc 128:10801–10807

Cheung CL, Rubinstein AI, Peterson EJ et al (2010) Steric and electrostatic complementarity in the assembly of two-dimensional virus arrays. Langmuir 26:3498–3505

Comellas-Aragonès M, Engelkamp H, Claessen VI et al (2007) A virus-based single-enzyme nanoreactor. Nat Nanotechnol 2:635–639

Comellas-Aragonès M, de la Escosura A, Dirks AT et al (2009) Controlled integration of polymers into viral capsids. Biomacromolecules 10:3141–3147

Dalsgaard K, Uttenthal Å, Jones TD et al (1997) Plant-derived vaccine protects target animals against a virus disease. Nat Biotechnol 15:248–252

De la Escosura A, Verwegen M, Sikkema FD et al (2008) Viral capsids as templates for the production of monodisperse Prussian blue nanoparticles. Chem Commun 1542–1544

Demir M and Stowell MHB (2002) A chemoselective biomolecular template for assembling diverse nanotubular materials. Nanotechnology 13:541–544

Dixit SK, Goicochea NL, Daniel MC et al (2006) Quantum dot encapsulation in viral capsids. Nano Lett 6:1993–1999

Douglas T and Young M (1998) Host–guest encapsulation of materials by assembled virus protein cages. Nature 393:152–55

Douglas T, Strable E, Willits D et al (2002) Protein engineering of a viral cage for constrained nanomaterials synthesis. Adv Mater 14:415–418

Dragnea B, Chen C, Kwak ES et al (2003) Gold nanoparticles as spectroscopic enhancers for in vitro studies on single viruses. J Am Chem Soc 125: 6374–6375

Dujardin E, Peet C, Stubbs G et al (2003) Organization of metallic nanoparticles using tobacco mosaic virus templates. Nano Lett 3:413–417

Evans DJ (2007) The bionanoscience of plant viruses: templates and synthons for new materials. J Mater Chem 18:3746–3754

Falkner JC, Turner ME, Bosworth JK et al (2005) Virus crystals as nanocomposite scaffolds. J Am Chem Soc 127:5274–5275

Fitchen J, Beachy RN, Hein MB (1995) Plant virus expressing hybrid coat protein with added murine epitope elicits autoantibody response. Vaccine 13:1051–1057

Fowler CE, Shenton W, Stubbs G et al (2001) Tobacco mosaic virus liquid crystals as templates for the interior design of silica mesophases and nanoparticles. Adv Mater 13:1266–1269

Franssen H, Goldbach R, Broekhuijsen M et al (1982) Expression of middle-component RNA of cowpea mosaic virus: in vitro generation of a precursor to both capsid proteins by a bottom-component RNA-encoded protease from infected cells. J Virol 41:8–17

Franzen S and Lommel SA (2009) Targeting cancer with 'smart bombs': equipping plant virus nanoparticles for a 'seek and destroy' mission. Nanomedicine 4:575–588

Gallie DR, Sleat DE, Watts JW et al (1987) In vivo uncoating and efficient expression of foreign mRNAs packaged in TMV-like particles. Science 236:1122–1124

Gillitzer E, Willits D, Young M et al (2002) Chemical modification of a viral cage for multivalent presentation. Chem Commun 2002:2390–2391

Gillitzer E, Suci P, Young M et al (2006) Controlled ligand display on a symmetrical protein-cage architecture through mixed assembly. Small 2:962–966

Hamamoto H, Sugiyama Y, Nakagawa N et al (1993) A new tobacco mosaic virus vector and its use for the systemic production of angiotensin-I-converting enzyme inhibitor in transgenic tobacco and tomato. Biotechnology 11:930–932

Hong V, Presolski SI, Ma C et al (2009) Analysis and optimisation of copper-catalysed azide-alkyne cycloadditions for bioconjugation. Angew Chem Int Ed 48:9879–9883

Huang X, Bronstein LM, Retrum J et al (2007) Self-assembled virus-like particles with magnetic cores. Nano Lett 7:2407–2416

Johnson JE, Lin T, Lomonossoff GP (1997) Presentation of heterologous peptides on plant viruses: genetics, structures and function. Annu Rev Phytopathol 35:67–86

Kaltgrad E, Gupta SS, Punna S et al (2007) Anti-carbohydrate antibodies elicited by polyvalent display on a viral scaffold. ChemBioChem 8:1455–1462

Kendall A, McDonald M, Bian W et al (2008) Structure of flexible filamentous plant viruses. J Virol 82:9546–9554

King DP, Montague N, Ebert K et al (2007) Development of a novel recombinant encapsidated RNA particle: evaluation as an internal control for diagnostic RT-PCR. J Virol Meth 146:218–225

Klem MT, Willits D, Young M et al (2003) 2-D array formation of genetically engineered viral cages on Au surfaces and imaging by atomic force microscopy. J Am Chem Soc 125: 10806–10807

Klem MT, Young M, Douglas T (2008) Biomimetic synthesis of ß-TiO_2 inside a viral capsid. J Mater Chem 18:3821–3823

Knez M, Bittner AM, Boes F et al (2003) Biotemplate synthesis of 3-nm nickel and cobalt nanowires. Nano Lett 3:1079–1082

Knez M, Sumser M, Bittner AM et al (2004a) Spatially selective nucleation of metal clusters on the tobacco mosaic virus. Adv Funct Mater 14:116–124

Knez M, Sumser MP, Bittner AM et al (2004b) Binding the tobacco mosaic virus to inorganic surfaces. Langmuir 20:441–447

Koo M, Bendahmane M, Lettieri GA et al (1999) Protective immunity against murine hepatitis virus (MHV) induced by intranasal or subcutaneous administration of hybrids of tobacco mosaic virus that carries an MHV epitope. Proc Natl Acad Sci USA 96:7774–7779

Koudelka KJ, Destito G, Plummer EM et al (2009). Endothelial targeting of cowpea mosaic virus (CPMV) via surface vimentin. PLoS Pathog. 5:e1000417

Kuncicky DM, Naik RR, Velev OD (2006) Rapid deposition and long-range alignment of nanocoatings and arrays of electrically conductive wires from tobacco mosaic virus. Small 2:1462–1466

Kwak M, Minten IJ, Anaya D-M (2010) Virus-like particles templated by DNA micelles: a general method for loading virus nanocarriers. J Am Chem Soc 132:7834–7835

Langeveld JP, Brennan FR, Martinez-Torrecuadrada JL et al (2001) Inactivated recombinant plant virus protects dogs from a lethal challenge with canine parvovirus. Vaccine 19:3661–3670

Lewis JD, Destito G, Zijlstra A et al (2006) Viral nanoparticles as tools for intravital vascular imaging. Nat Med 12:354–360

Lico C, Capuano F, Renzone G et al (2006) Peptide display on potato virus X: molecular features of the coat protein-fused peptide affecting cell-to-cell and phloem movement of chimeric virus particles. J Gen Virol 87:3103–3112

Lico C, Mancinia C, Italiani P et al (2009) Plant-produced potato virus X chimeric particles displaying an influenza virus-derived peptide activate specific CD8+ T cells in mice. Vaccine 27:5069–5076

Liepold L, Anderson S, Willits D et al (2007) Viral capsids as MRI contrast agents. Magn Reson Med 58:871–879

Lin T, Johnson JE (2003) Structures of picorna-like plant viruses: implications and applications. Adv Virus Res 62:167–239

Lin T, Porta, C, Lomonossoff G et al (1996) Structure-based design of peptide presentation on a viral surface: the crystal structure of a plant/animal virus chimaera at 2.8Å resolution. Folding and Design 1:179–187

Lomonossoff GP (2005) Antigen delivery systems: Use of recombinant plant viruses. In: Mestecky J, Bienenstock J, Lamm ME et al (eds) Mucosal Immunology, 3rd edn. Elsevier, pp 1061–1072

Lomonossoff GP, Hamilton WDO (1999) Cowpea mosaic virus-based vaccines. Curr Top Microbiol Immunol 240:177–189
Lomonossoff GP, Johnson JE (1995) Eukaryotic viral expression systems for polypeptides. Semin Virol 6:257–267
Lomonossoff GP, Johnson JE (1996) Use of macromolecular assemblies as expression systems for peptides and synthetic vaccines. Curr Opin Struct Biol 6:176–182
Loo L, Guenther RH, Basnayake VR et al (2006) Controlled encapsidation of gold nanoparticles by a viral protein shell. J Am Chem Soc 128:4502–4503
Loo L, Guenther RH, Lommel SA et al (2007) Encapsidation of nanoparticies by red clover necrotic mosaic virus. J Am Chem Soc 129:11111–11117
Lvov Y, Haas H, Decher G et al (1994) Successive deposition of alternate layers of polyelectrolytes and a charged virus. Langmuir 10:4232–4236
Marusic C, Rizza P, Lattanzi L et al (2001) Chimeric plant virus particles as immunogens for inducing murine and human immune responses against Human immunodeficiency virus type 1. J Virol 75:8434–8439
McCormick A, Palmer KE (2008) Genetically engineered tobacco mosaic virus as nanoparticle vaccines. Expert Rev Vaccines 7:33–41
Medintz IL, Sapsford KE, Konnert JH et al (2005) Decoration of discretely immobilized cowpea mosaic virus with luminescent quantum dots. Langmuir 21:5501–5510
Meunier S, Strable E, Finn MG (2004) Crosslinking of and coupling to viral capsid proteins by tyrosine oxidation. Chem Biol 11:319–326
Mukherjee S, Pfeifer CM, Johnson JM et al (2006) Redirecting the coat protein of a spherical virus to assemble into tubular nanostructures. J Am Chem Soc 128:2538–2539
Ochoa W, Chatterji A, Lin T et al (2006) Generation and structural analysis of reactive empty particles derived from an icosahedral virus. Chem Biol 13:771–778
Phelps JP, Dang N, Rasochova L (2007) Inactivation and purification of cowpea mosaic virus-like particles displaying peptide antigens from *Bacillus anthracis*. J Virol Meth 141:146–153
Porta C, Spall VE, Loveland J et al (1994) Development of cowpea mosaic virus as a high-yielding system for the presentation of foreign peptides. Virology 202:949–955
Porta C, Spall VE, Lin T et al (1996) The development of cowpea mosaic virus as a potential source of novel vaccines. Intervirol 39:79–84
Porta C, Spall VE, Findlay KC et al (2003) Cowpea mosaic virus-based chimaeras. Effects of inserted peptides on the phenotype, host-range and transmissibility of the modified viruses. Virology 310:50–63
Prasuhn DE, Yeh RM, Obenaus A et al (2007) Viral MRI contrast agents: coordination of Gd by native virions and attachment of Gd complexes by azide-alkyne cycloaddition. Chem Commun, pp 1269–1271
Rae C, Koudelka KJ, Destito G et al (2008) Chemical addressability of ultraviolet-inactivated viral nanoparticles (VNPs). PLoS ONE 3:e3315
Raja KS, Wang Q, Gonzalez MJ et al (2003) Hybrid virus-polymer materials. 1. Synthesis and properties of PEG-decorated cowpea mosaic virus. Biomacromolecules 4:472–476
Ren Y. Wong S-M, Lim L-Y (2006) In vitro-reassembled plant virus-like particles for loading of polyacids. J Gen Virol 87:2749–2754
Ren Y, Wong S-M, Lim L-Y. (2007) Folic acid-conjugated protein cages of a plant virus: a novel delivery platform for doxorubicin. Bioconjug Chem 18:836–843
Royston ES, Ghosh A, Kofinas P et al (2008) Self-assembly of virus structured high surface area nanomaterials and their application as battery electrodes. Langmuir 24:906–912
Royston ES, Brown AD, Harris MT et al (2009) Preparation of silica stabilized tobacco mosaic virus templates for the production of metal and layered nanoparticles. J Colloid Interface Sci 332:402–407
Sainsbury F, Cañizares MC, Lomonossoff GP (2010) Cowpea mosaic virus: the plant virus-based biotechnology workhorse. Annu Rev Phytopathol 48:437–455
Santa Cruz S, Chapman S, Roberts AG et al (1996) Assembly and movement of a plant virus carrying a green fluorescent protein overcoat. Proc Natl Acad Sci USA 93:6286–6290

Saunders K, Sainsbury, F, Lomonossoff GP (2009) Efficient generation of cowpea mosaic virus empty virus-like particles by the proteolytic processing of precursors in insect cells and plants. Virology 393:329–337

Schlick TL, Ding ZB, Kovacs EW et al (2005) Dual-surface modification of the tobacco mosaic virus. J Am Chem Soc 127:3718–3723

Sen Gupta S, Kuzelka J, Singh P et al (2005a) Accelerated bioorthogonal conjugation: a practical method for the ligation of diverse functional molecules to a polyvalent virus scaffold. Bioconjug Chem 16:1572–1579

Sen Gupta S, Raja KS, Kaltgrad E et al (2005b) Virus-glycopolymer conjugates by copper(I) catalysis of atom transfer radical polymerization and azide-alkyne cycloaddition. Chem Commun 14:4315–4317

Shah SN, Steinmetz NF, Aljabali AAA et al (2009) Environmentally benign synthesis of virus-templated, monodisperse, iron-platinum nanoparticles. Dalton Trans, pp 8479–8480

Shenton W, Douglas T, Young M et al (1999) Inorganic–organic nanotube composites from template mineralization of tobacco mosaic virus. Adv Mater 11:253–256

Singh P, Prasuhn D, Yeh RM et al (2007) Bio-distribution, toxicity and pathology of cowpea mosaic virus nanoparticles in vivo. J Control Release 120:41–50

Sit TL, Vaewhongs AA, Lommel SA (1998) RNA-mediated trans-activation of transcription from a viral RNA. Science 281:829–832

Smith JC, Lee KB, Wang Q et al (2003) Nanopatterning the chemospecific immobilization of cowpea mosaic virus capsid. Nano Lett 3:883–886

Smith ML, Lindbo JA, Dillard-Telm S et al (2006) Modified tobacco mosaic virus particles as scaffolds for display of protein antigens for vaccine applications. Virology 348:475–488

Smith ML, Corbo T, Bernales J et al (2007) Assembly of trans-encapsidated recombinant viral vectors engineered from tobacco mosaic virus and semliki forest virus and their evaluation as immunogens. Virology 358:321–333

Smith ML, Fitzmaurice WP, Turpen TH et al (2009) Display of peptides on the surface of tobacco mosaic virus particles. Curr Top Microbiol Immunol 332:13–31

Smolenska L, Roberts IM, Learmonth D et al (1998) Production of a functional single chain antibody attached to the surface of a plant virus. FEBS Lett 441:379–382

Speir JA, Munshi S, Wang G et al (1995) Structures of the native and swollen forms of cowpea chlorotic mottle virus determined by X-ray crystallography and cryo-electron microscopy. Structure 3: 63–78

Steinmetz NF, Evans DJ (2007) Utilisation of plant viruses in bionanotechnology. Org Biomol Chem 5:2891–2902

Steinmetz NF, Manchester M (2009) PEGylated viral nanoparticles for biomedicine: the impact of PEG chain length on VNP cell interactions in vitro and ex vivo. Biomacromolecules 13:10784–10792

Steinmetz NF, Calder G, Lomonossoff GP et al (2006a) Plant viral capsids as nanobuilding blocks: construction of arrays on solid supports. Langmuir 22:10032–10037

Steinmetz NF, Lomonossoff GP, Evans DJ (2006b) Decoration of cowpea mosaic virus with multiple, redox-active, organometallic complexes. Small 2:530–533

Steinmetz NF, Lomonossoff GP, Evans DJ (2006c) Cowpea mosaic virus for material fabrication: addressable carboxylate groups on a programmable nanoscaffold. Langmuir 22:3488–3490

Steinmetz NF, Evans DJ, Lomonossoff GP (2007) Chemical introduction of reactive thiols into a viral nanoscaffold: a method which avoids virus aggregation. ChemBioChem 8:1131–1136

Steinmetz NF, Bock E, Richter RP et al (2008a) Assembly of multilayer arrays of viral nanoparticles via biospecific recognition: a quartz crystal microbalance with dissipation monitoring study. Biomacromolecules 9:456–462

Steinmetz NF, Findlay KC, Noel TR et al (2008b) Layer-by-layer assembly of viral nanoparticles and polyelectrolytes: The film architecture is different for spheres versus rods. ChemBioChem 9:1662–1670

Steinmetz NF, Lin T, Lomonossoff GP et al (2009a) Structure-based engineering of an icosahedral virus for nanomedicine and nanotechnology. Curr Top Microbiol Immunol 327:23–58
Steinmetz NF, Shah SN, Barclay JE et al (2009b) Virus-templated silica nanoparticles. Small 5: 813–816
Steinmetz NF, Hong V, Spoerke ED et al (2009c) Buckyballs meet viral nanoparticles: candidates for biomedicine. J Am Chem Soc 131:17093–17095
Steinmetz NF, Mertens ME, Taurog RE et al (2010) Potato virus X as a novel platform for potential biomedical applications. Nano Lett10:305–312
Strable E, Finn MG (2009) Chemical modification of viruses and virus-like particles. Curr Top Microbiol Immunol 332:1–21
Suci PA, Klem MT, Douglas T et al (2005) Influence of electrostatic interactions on the surface adsorption of a viral protein cage. Langmuir 21:8686–8893
Suci PA, Klem MT, Arce FT et al (2006) Assembly of multilayer films incorporating a viral protein cage architecture. Langmuir 22: 8891–8896
Suci PA, Berglund DL, Liepold L et al (2007a) High-density targeting of a viral multifunctional nanoplatform to a pathogenic, biofilm-forming bacterium. Chem Biol 14:387–398
Suci PA, Varpness Z, Gillitzer E et al (2007b) Targeting and photodynamic killing of a microbial pathogen using protein cage architectures functionalized with a photosensitizer. Langmuir 23:12280–12286
Sugiyama Y, Hamamoto H, Takemoto S et al (1995) Systemic production of foreign peptides on the particle surface of Tobacco mosaic virus. FEBS Lett 359:47–250
Sun J, DuFort C, Daniel MC et al (2007) Core-controlled polymorphism in virus-like particles. Proc Natl Acad Sci USA 104:1354–1359
Tang J, Johnson JM, Dryden KA et al (2006) The role of subunit hinges and molecular "switches" in the control of viral capsid polymorphism. J Struct Biol 154:59–67
Taylor KM, Lin T, Porta C. et al (2000) Influence of three-dimensional structure on the immunogenicity of a peptide expressed on the surface of a plant virus. J Mol Recognit 13: 71–82
Tsukamoto R, Muraoka M, Seki M et al (2007) Synthesis of CoPt and $FePt_3$ nanowires using the central channel of Tobacco mosaic virus as a biotemplate. Chem Mater 19:2389–2391
Turpen TH, Reinl SJ, Charoenvit Y et al (1995) Malarial epitopes expressed on the surface of recombinant Tobacco mosaic virus. Biotechnology 13:53–57
Uhde-Holzem K, Schlösser V, Viazov S et al (2010) Immunogenic properties of chimeric Potato virus X particles displaying the Hepatitis C virus hypervariable region I peptide R9. J Virol Meth 166:12–20
Usha R, Rohll JB, Spall VE et al (1993) Expression of an animal virus antigenic site on the surface of a plant virus particle. Virology 197:366–374
Wang Q, Kaltgrad E, Lin T et al (2002a) Natural supramolecular building blocks: wild-type Cowpea mosaic virus. Chem Biol 9:805–811
Wang Q, Lin T, Tang L et al (2002b) Icosahedral virus particles as addressable nanoscale building blocks. Angew Chem Int Ed 41:459–462
Wang Q, Lin T, Johnson JE et al (2002c) Natural supramolecular building blocks: cysteine-added mutants of cowpea mosaic virus. Chem Biol 9:813–819
Werner S, Marillonnet S, Hause G et al (2006) Immunoabsorbent nanoparticles based on a tobamovirus displaying protein A. Proc Natl Acad Sci USA 103:17678–17683
Yi HM, Nisar S, Lee SY et al (2005) Patterned assembly of genetically modified viral nanotemplates via nucleic acid hybridization. Nano Lett 5:1931–1936
Yi HM, Rubloff GW, Culver JN (2007) TMV microarrays: hybridization-based assembly of DNA-programmed viral nanotemplates. Langmuir 23:2663–2667
Young M, Willits D, Uchida M et al (2008) Plant viruses as biotemplates for materials and their use in nanotechnology. Annu Rev Phytopathol 46:361–384

Milestones in the Development and Applications of Plant Virus Vector as Gene Silencing Platforms

Christophe Lacomme

Abstract One of the main post-genomics challenges facing scientists remains the identification of gene function in a large number of plant species. Plant viruses offer great potential in linking genes to phenotypes through epigenetic expression or knockdown of selected genes. The past decade has seen the development and ever increasing applications of a gene knockdown technique termed virus-induced gene silencing (VIGS). VIGS recapitulates an RNA-mediated antiviral defense mechanism, mediating a homology-based post-transcriptional degradation of selected plant RNAs, leading to a loss-of-function phenotype. Due to its rapidity and increasing number of virus vectors developed as gene silencing platforms, VIGS has become a powerful technology to determine the function of genes in an increasing number of crop species, where the routinely available transgenesis or mutagenesis approaches are often not amenable to large genomes and complex genetic backgrounds.

Contents

C. Lacomme (✉)
Virology and Zoology Section,
Science and Advice for Scottish Agriculture (SASA),
Roddinglaw Road, Edinburgh, EH12 9FJ, UK
e-mail: Christophe.Lacomme@sasa.gsi.gov.uk

Current Topics in Microbiology and Immunology (2011) 375: 89–105
DOI: 10.1007/82_2011_186

Published Online: 28 October 2011

1 Discovery of Virus-Induced Gene Silencing (VIGS) Phenomenon

In the early 1990s, studies on pathogen-derived resistance uncovered an unsuspected mechanism where coat protein (CP) transgene-mediated resistance to infecting viruses was mediated by a non-coding transgenic CP RNA (Lindbo and Dougherty 1992; Lindbo et al. 1993). The key finding of this seminal work was that the viral RNA was targeted for degradation in a sequence-specific manner. Further, other laboratories demonstrated that this phenomenon was not restricted to transgenic plants. Indeed a similar phenomenon was found to occur during natural viral infections, suggesting the involvement of an endogenous RNA-mediated defense mechanism affecting both DNA and RNA viruses where plants "recover" from virus infection (Al-Kaff et al. 1998; Ratcliff et al. 1997; Matthews 1991). Recovery is defined by the formation of asymptomatic tissue at the top of a virus-infected plant (Matthews 1991). Recovered tissues are free of, or contain very low amounts of, viral RNA (Matthews 1991; Moore et al. 2001; Mlotshwa et al. 2008). Collectively, these data helped to understand the nature of the recovery phenomenon observed in response to virus infection as being reminiscent of other phenomena termed cosuppression or post-transcriptional gene silencing (PTGS) where viral and transgene RNA undergo post-transcriptional degradation.

"Virus-induced gene silencing" was first coined by van Kammen (1997). In this review entitled "Virus-induced gene silencing in infected and transgenic plants", VIGS was initially used to describe the RNA-mediated defense mechanism in transgenic plants expressing viral sequences. The resistance response observed to sequence-related (homologous) viruses was expressed in the form of recovery from the initial infection.

Having demonstrated that transgenic plants could direct RNA degradation to infecting viruses, it was not until 1995 when scientists established that a genetically modified viral expression vector harboring a plant cDNA could in turn silence the corresponding endogenous plant gene (Kumagai et al. 1995). Virus-induced gene silencing was illustrated by the silencing of a *phytoene desaturase* (*pds*) gene leading to the photobleaching of the silenced leaves, appearing white as a consequence of chlorophyll pigments degradation (Fig. 1a). This early landmark of virus-induced gene silencing (VIGS) exemplified the potential of virus-derived expression vectors as powerful tools for rapid gene knock-down for gene function analysis; and pioneered the development and use of VIGS as a functional genomic platform in plants.

2 Biology of Virus-Induced Gene Silencing

2.1 The RNA Silencing Machinery in Plants

During natural viral infection, viruses are targets of the RNA silencing machinery, triggering the degradation of their genome. RNA-mediated defense mechanism or RNA silencing was then found to shape virus infections in many ways. Synergism between viruses in mixed infections (for example between potex- and poty-viruses, Mlotshwa et al. 2008) and meristem exclusion of some viruses (Schwach et al. 2005) represent different manifestations of the endogenous antiviral RNA silencing response. Synergy involves exacerbation of viral symptoms during mixed virus infection as a consequence of virus-mediated suppression of RNA silencing. Meristem exclusion is a manifestation of plants using RNA silencing to control virus movement and distribution within the plant (Mlotshwa et al. 2008; Schwach et al. 2005). The discovery of virus-encoded silencing suppressors provided further insight into the dynamics of plant–virus interactions, illustrating the fact that viruses have co-evolved diverse counter-defensive strategies to overcome the host RNA silencing machinery (Voinnet et al. 1999).

Physical manifestations of the antiviral silencing response, which leads to recovery from viral infection, can range from the appearance of punctate dark green islands of virus-free tissues among infected tissues or asymptomatic new emerging leaves with a significantly reduced virus titer or in some cases complete elimination (Matthews 1991; Moore et al. 2001; Ratcliff et al. 1997; Baulcombe 2004). Depending on the virus-host interaction dynamics, the manifestation and the extent of the RNA silencing response can be diverse, ranging from small areas of an infected leaf (Moore et al. 2001) to entire leaves above the inoculated leaf (Ratcliff et al. 1997; Baulcombe 2004; Mlotshwa et al. 2008). In comparison with other resistance mechanisms such as the hypersensitive response (HR) (a form of localized cell death leading to the confinement of the pathogen at its infection site), RNA silencing is a relatively slow process requiring several days to weeks before being observed in younger upper uninoculated leaves and is dependent on environmental conditions (Szittya et al. 2003; Tuttle et al. 2008).

Genetic analysis of RNA silencing revealed that in plants, the RNA silencing pathway is not only associated with resistance to viruses but also plays an important role in plant development, maintenance of genome integrity and regulation of gene expression (Brodersen and Voinnet 2006). RNA silencing has recently been associated with other plant pathogens including bacteria (Navarro et al. 2006; Katiyar-Agarwal et al. 2006). RNA silencing underlies other phenomena observed

in plants (such as co-suppression and post-transcriptional gene silencing) and in other kingdoms such as quelling in fungi (Fulci and Macino 2007) or RNA interference (RNAi) in nematodes (Grishok 2005). Due to its omnipresence, RNA silencing emerged as an ancient mode of genome expression regulation at the post-transcriptional and transcriptional level (Brodersen and Voinnet 2006).

◀**Fig. 1** Examples of virus-induced gene silencing and phenotypes generated. **a** Photobleaching induced by Tobacco rattle virus (TRV.PDS) and Potato virus X (PVX.PDS) as a consequence of *phytoene desaturase* (*pds*) silencing in *N. benthamiana*. Note the recovery of *pds* silencing occurring with the PVX.PDS VIGS construct (*right*) where recovered green tissues appears on young emerging leaves (by 12 days post inoculation) as opposed to TRV.PDS silencing. **b** Examples of silencing of endogenous genes in barley using Barley stripe mosaic virus (BSMV). Mock, uninfected leaves; BSMV.PDS, BSMV VIGS vector silencing barley *pds* gene (photobleached leaves appeared *white*); BSMV.GFP, control leaves infected by BSMV construct expressing GFP as a control of virus infection; BSMV.Sulfur, BSMV construct silencing the barley *Magnesium chelatase* gene (*ChlH* or *sulfur*), silenced leaves display a light yellow coloration; BSMV.LLS1, silencing or *Lethal Leaf Spot* 1 generating runaway cell death and necrosis on the silenced leaves phenocopying *acd1* (*accelerated cell death 1*) Arabidopsis mutants. **c** Example of systemic silencing of genes involved in the carotenoid biosynthetic pathway in potato. PVX-mediated VIGS of *pds* in potato leaves (*upper left panel*), and in tubers (biochemical monitoring of *cis*- and *trans*-phytoene accumulation increased in silenced tissues, upper *right* panel) (Faivre-Rampant et al. 2004). PVX-mediated VIGS of *zeaxanthin epoxidase* (*zep*) gene (PVX.ZEP). Acetone extract from *zep* silenced lyophilized tubers appears *yellow* as a consequence of *zep* silencing (*bottom panel*) (PVX.GFP, virus-infected control tubers). The silenced genes from the carotenoid biosynthetic pathway are marked with a cross. A color version is available in the online version

The initial steps of RNA silencing require the generation of double-stranded (ds)RNA that are perceived as aberrant RNA. DsRNA is then processed into small interfering (si)RNAs of a size ranging between 21 and 25 nt in length by RNAse III-like enzymes (Dicer-Like, DCL) dsRNA processing enzymes. These siRNAs are in turn incorporated into a RNA-induced silencing complex (RISC) that will recognize and hybridize to complementary (or partially complementary) target RNA leading to its degradation (Dunoyer and Voinnet 2005; Brodersen and Voinnet 2006). The presence of loaded RISC-complex with viral siRNA will guide viral RNA degradation in subsequent infection by the same virus or a related virus bearing sufficient sequence homology and contribute to the maintenance of the silencing response (Pantaleo et al. 2007). The discovery that siRNAs act as templates in guiding the RISC-complex to degrade homologous RNA demonstrates the short size of target gene sequence information required to trigger gene silencing (Hamilton and Baulcombe 1999; Thomas et al. 2001).

2.2 Origin of Viral Double-Stranded(ds) RNA

As previously stated, the initial step of RNA silencing involves the production and degradation of aberrant RNAs. Aberrant RNAs can be of different nature and origins (Herr et al. 2006; Wypijewski et al. 2009). The core enzymes of the silencing machinery such as DCL, Argonautes and accessory RNA-dependant RNA polymerases (RDR) have been shown to be involved in different RNA silencing pathways associated with the degradation of diverse aberrant RNAs (Brodersen and Voinnet 2006; Herr et al. 2006; Wypijewski et al. 2009). One common feature of the RNA silencing response is the direct or indirect generation of dsRNA, a step that appears to be universally necessary to engage RNA silencing pathways. During virus infection, it is generally believed that for positive (+)-stranded RNA viruses, viral dsRNAs are generated by the pairing between replicative intermediates between the positive (+) and negative (−) viral RNA strand. However, sequence analysis of siRNA populations during several virus-plant infections have identified a higher proportion of siRNAs of (+) strand polarity as opposed to (−) strand polarity, arguing against the intermolecular pairing between replicative intermediates as a potential origin of viral dsRNA. This suggest that, at least for several virus species, viral dsRNA formation is unlikely to originate from replicative intermediates of the (+) and (−) strand. Instead, viral dsRNA may predominantly arise from secondary structure originating from the (+) strand, for some viruses such as Tobacco mosaic virus (TMV), Potato virus X (PVX), Cymbidium ringspot mosaic virus (CyMRSV), Tobacco rattle virus (TRV) and Turnip crinkle carmovirus (TCV) (Molnár et al. 2005; Ho et al. 2006; Donaire et al. 2008). Alternatively, the involvement of multiple RDRs using the viral (+) strand as a template to synthesize a complementary (−) strand to generate dsRNA in the process, could explain the preferential accumulation of siRNAs in several regions of the viral genome (Donaire et al. 2008).

2.3 Dynamics of the Virus-Induced Silencing Response

Plant–virus interactions are dynamic and both viruses and plants have each co-evolved to fend off defense mechanisms from the host and pathogenicity determinants from the virus (Dunoyer and Voinnet 2005). With the discovery of viral-encoded silencing suppressors and their role in regulating many aspects of plant–virus interactions (such as virus replication, local and systemic movement, symptom development and synergism between viruses), RNA silencing emerged as a key mechanism shaping virus accumulation and distribution within the plant (Dunoyer and Voinnet 2005). During virus replication, local and systemic movement, large amounts of genomic (and subgenomic) viral RNA will be generated and, in turn, will be targeted for degradation by the silencing machinery. In the past decade, surveys of several viruses have identified silencing suppressors that interfere in different ways with the silencing response to protect the viral genome against RNA degradation (Voinnet et al. 1999; Dunoyer and Voinnet 2005). It is generally believed that most, if not all, viruses harbor silencing suppressors influencing their pathogenicity. Virus-plant interactions, and therefore the viral-induced silencing response, will be strongly influenced by the capacity of the host to degrade viral genome and the ability of the virus to escape or suppress the RNA silencing response to move systemically in the plant.

Consequently, the dynamic of the silencing response varies between virus and host species combinations (Robertson 2004). For example, in *N. benthamiana* using a TMV VIGS vector it was demonstrated that the silencing response observed during *ChlH* silencing, despite fluctuating (fading and reappearing in emerging leaves), could still be observed for long period of time (Hiriart et al. 2003). A sustained silencing response was observed not only in *N. benthamiana* but also in crop species including potato (Fig. 1c, Faivre-Rampant et al. 2004), and barley (Fig. 1b, Hein et al. 2005) using PVX and BSMV VIGS vectors, respectively. In the case of PVX VIGS vector, silencing of *pds* in *N. benthamiana* was originally strong but faded rapidly, as opposed to VIGS triggered by a TRV VIGS vector (Fig. 1a). In contrast, the dynamics of *pds* silencing mediated by PVX appeared to be different in potato in comparison to *N. benthamiana*, with photobleaching more localised to the vicinity of veins (Fig. 1a–c) but remaining stable for longer period of time to ultimately being observed in mature tubers months after the initial infection (Fig. 1c, Faivre-Rampant et al. 2004). This suggests that, at least in the case of endogenous gene silencing by RNA viruses, complete elimination of the VIGS vector does not generally occur, leading to a reduced but significant accumulation of the VIGS vector virus, promoting a durable silencing response with various degrees of intensity.

2.4 Parameters Affecting the VIGS Response

The outcome of the silencing response depends largely on the dynamics of virus-host interactions and optimal environmental conditions compatible with plant growth and virus accumulation (Burch-Smith et al. 2004). Among these, temperature is known to affect the RNA silencing response (Szittya et al. 2003; Tuttle et al. 2008). Differences in susceptibility to virus infection and movement are often observed between cultivars or accessions of the same plant species (Hein et al. 2005). Therefore careful optimization of VIGS response requires the assessment of different cultivars and a range of experimental conditions to obtain a robust silencing response.

N. benthamiana, the model plant for plant virologist due to its propensity to tolerate high level of accumulation of a wide range of viruses, has become a model for VIGS development and phenomics (Goodin et al. 2008). A previous report identified a non-functional antiviral RDR1 variant in *N. benthamiana* (Yang et al. 2004). This lack of the salicylic acid- and virus-inducible RNA-dependant RNA polymerase could explain the increased susceptibility to viruses and indirectly the predisposition of *N. benthamiana* to support a robust VIGS response (Yang et al. 2004).

The nature of the silenced target gene has a strong influence on the outcome of the silencing response. Non-cell autonomous spreading of the silencing response is observed during transgene silencing concomitantly with exacerbed RNA degradation of the trigger recombinant virus; as opposed to the silencing of endogenous genes. During transgene silencing, non-cell autonomous silencing can be initiated locally and spread systemically via an RDR6-dependant pathway in the absence of systemic movement of the silencing construct (Voinnet 2005). Contrastingly, the characteristics of the silencing response for endogenous genes (such as *pds*) induced by viruses is tightly associated with cells supporting viral replication. This explains the phenotype associated with VIGS of endogenous genes that are observed mainly in tissues supporting virus replication (Himber et al. 2003).

Intrinsic characteristics of the gene to be silenced such as, protein turnover and tissular expression (for example in meristematic tissues, from which most viruses are naturally excluded), will affect the outcome of the silencing response. Moreover, it was reported that DCL activity has a strong bias for high GC content templates (Ho et al. 2007). Selection of suitable insert with specific GC content could potentially influence the robustness of the silencing response.

Numerous studies have reported the propensity of virus-derived vectors to recombine when harboring heterologous sequences and consequently selecting recombinant virus with shorter cDNA. Analysis of recombinant genomes suggested that for BSMV, a portion of approximately 100-nt in length was retained and displayed silencing for longer periods of time as opposed to larger fragments that were more rapidly deleted (Bruun-Rasmussen et al. 2007). Due to the limited size of gene fragments required to generate VIGS in this system, a sustained silencing response can still be observed after several weeks following virus infection (Hein et al. 2005; Lacomme et al. 2003), persisting through seed tissues and still being observed in the next generation of plants (Bruun-Rasmussen et al. 2007).

3 Development and Application of VIGS for Gene Function Characterization

Prior to VIGS, genetically engineered plant viruses were used for gene function characterization through overexpression of foreign cDNAs (Lacomme et al. 2001). This led to several applications, for example, in the field of plant pathogen interactions where overexpression of non-endogenous full-length cDNAs (Rommens et al. 1995; Lacomme and Santa Cruz 1999; Takken et al. 2000) or a plant-derived cDNA library (Karrer et al. 1998) provided valuable informations on the molecular mechanisms associated with plant resistance in response to pathogen challenge.

These studies illustrated the flexibility and potential of virus vectors to accelerate gene function studies and the development of high-throughput gene function screens. However, one of the main limitations of virus expression vectors is the length of insert size tolerated in their genome. Increasing viral genome size leads to recombination events, selection of fitter recombinant virus harboring smaller sized inserts and in most of the cases a rapid lack of functionality of the expressed protein (Lacomme et al. 2001). This inconvenience has restricted the use of some virus groups, in particular icosahedral viruses as opposed to rod-shaped viruses, for full-length expression of proteins, where packaging constraints for icosahedral viruses are likely to impose an even stronger selective pressure in favor of recombinants with a size closer to a wild-type genome. In the case of VIGS, the insert size required to generate a silencing response is much smaller, on average between 100 and 500 nt in length (Burch-Smith et al. 2004) down to 21 nt in some cases (Thomas et al. 2001). The relatively small insert size requirement for generation of a robust silencing response allows the the use of virus expression systems that were previously hindered by this limitation for full-length cDNA expression.

3.1 Viruses Used for Gene Silencing

As previously stated, the first example of *bona fide* virus-induced gene silencing was illustrated by the silencing of *phytoene desaturase* gene using a TMV expression vector (Kumagai et al. 1995). This inhibition of endogenous gene expression by a cytoplasmic episomal RNA viral vector demonstrated that viral RNA can be used to re-direct the endogenous RNA silencing pathway to knock-down host gene expression.

Since then, a vast range of viruses have been developed as VIGS vectors (Table 1). These include positive-stranded RNA viruses group (examples of genera and representative viruses whose VIGS vector are derived from are tobravirus: TRV, potexvirus: PVX, hordeivirus: Barley stripe mosaic virus, BSMV, comovirus: Bean pod mottle virus, BPMV), single-stranded DNA viruses group (Begomoviruses: Tomato golden mosaic virus, TGMV, African cassava mosaic virus, ACMV, Cabbage leaf curl virus, CaLCuV) and other viral genetic elements such as satellite RNA satellite tobacco mosaic virus (STMV) (Table 1, Robertson 2004; Burch-Smith et al. 2004).

3.2 Forward and Reverse Genetic Screens

The demonstration of the applicability of VIGS as a high-throughput functional genomic approach was reported by Lu et al. (2003) by the identification of genes associated with the HR development during pathogen resistance. In this forward genetic screen, a normalized cDNA library was cloned into a PVX VIGS vector, followed by the screening of nearly 5,000 individual VIGS construct. The authors identified three Heat Shock Protein 90 clones preventing HR-associated cell death in silenced tissues (Lu et al. 2003). Another example was provided with sense and antisense TMV and BSMV libraries representing 11,200 and 1,100 unique cDNA clones arrayed in 96-well plate format, followed by high-throughput inoculation of transcripts of *N. benthamiana* or barley, respectively (Fitzmaurice et al. 2002). The downstream analysis of gain or loss of function screens revealed distinctive phenotypes for $\sim$5% of transfected plants. In addition, such high-throughput "phenomic" platform could be coupled with biochemical profiling of a range of

Table 1 Selected examples of VIGS vectors and their main characteristics

Virus group	Genus	Virus name	Host range	Tissue silenced	Applications	References
(+) ssRNA	Tobravirus	Tobacco rattle virus (TRV)	Solanacae, papaveracae Brassicacae	Leaves, roots, fruits, seeds, flowers, meristems	Development Plant–pathogen interactions	Liu et al. (2002) Wege et al. (2007) Valentine et al. (2004)
		Pea early browning virus (PEBV)	Fabacae	Leaves, roots	Plant-symbiont interactions	Constantin et al. (2008)
	Potexvirus	Potato virus X (PVX)	Solanacae	Leaves, tubers	Development Plant–pathogen interactions	Lu et al. (2003) Faivre-Rampant et al. (2004)
	Hordeivirus	Barley stripe mosaic virus (BSMV)	Poacae	Leaves, seeds	Development Plant–pathogen interactions	Holzberg et al. (2002) Hein et al. (2005) Scofield et al. (2005)
	Bromovirus	Brome mosaic virus (BMV)	Poacae	Leaves	Development Plant–pathogen interactions	Ding et al. (2006)
	Cheravirus	Apple latent spherical virus (ALSV)	Brassicacae, fabacae Cucurbitacae, solanacae	Leaves	Development	Igarashi et al. (2009)
	RNA satellite virus	Satellite tobacco mosaic virus (STMV)	Solanacae (*N. tabacum*)	Leaves, roots, flowers	Development	Gosselé et al. (2002)
ssDNA	Begomovirus	Tomato golden mosaic virus (TGMV)	Solanacae	Leaves, meristems	Development	Kjemtrup et al. (1998) Peele et al. (2001)
		Cabbage leaf curl virus (CaLCuV)	Brassicacae	Leaves	Development	Turnage et al. (2002)
		African cassava mosaic virus (ACMV)	Euphorbiacae	Leaves	Development	Fofana et al. (2004)

selected metabolites. These technical "tour de force" illustrate well the potential of virus vectors for high-throughput phenomics. Importantly they offer the possibility of linking genes to traits in identifying the function of genes causing developmental lethality that cannot be harnessed through transgenesis or mutagenesis.

The large proportion of VIGS-based functional characterizations are based on reverse genetic screens, where specific cDNAs corresponding to selected target genes are cloned into a VIGS vector to analyze the effect of gene downregulation on a specific trait (Table 1, Fig. 1a–c) as reported in the following section.

3.3 Applications of VIGS

Due to its natural propensity to tolerate relatively high level of virus accumulation *Nicotiana benthamiana* has rapidly become the model plant for VIGS-based studies. Other characteristics make *N. benthamiana* a first choice for in reverse genetics approaches (Goodin et al. 2008). Importantly, the ease of Agrobacteria-mediated transient expression makes *N. benthamiana* a well-suited host to study a vast array of molecular and cellular events associated with plant-pathogen interactions. Moreover, it appeared that common patterns of resistance responses (host, non-host and basal resistance) can be applied in *N. benthamiana* for a wide range of pathogens and pests (bacteria, fungi, oomycetes and nematodes). Agrobacterium-mediated expression of pathogen and host effectors have been shown to reproduce molecular mechanisms associated with resistance responses for a large set of resistance genes from heterologous plant–pathogen interactions in *N. benthamiana* (Gilroy et al. 2007; Goodin et al. 2008). The design of screens combining transient Agrobacterium-mediated expression with VIGS in *N. benthamiana* has contributed to the understanding of key molecular events in plant-pathogen interactions (Table 1, Burch-Smith et al. 2004).

Over the past decades, the applications of VIGS have steadily increased. The demonstration that VIGS can be extended to an increasing number of viruses with very different host ranges (Table 1), demonstrated the power of this approach to rapidly link genes with phenotypes in a large number of plants. VIGS has been reported in many plant species including Solanacae (tobacco, tomato, potato), Fabacae (soybean, alfalfa, pea, bean), Brassicacae (*Arabidopsis thaliana*), Poacae (barley, wheat, rice, maize), Papaveracae (poppy) and Euphorbiacae (cassava); in a variety of tissues and organs (leaf, tubers, roots, fruits and seed tissues) (Table 1). For most of these species, VIGS is the only approach available to undertake gene function analysis due to their genome complexity and poor transformation efficiency. Apart from plant–pathogen interactions studies, VIGS-based screens have been designed to study development, metabolic pathways, RNA silencing, symbiosis and tuber-associated traits (Table 1).

3.4 Optimization and Recent Development of VIGS Systems

As for all plant–virus interactions, environmental conditions strongly influence virus replication, movement and ultimately the silencing response. Efforts have been made in different VIGS systems to identify the optimal conditions to generate a robust VIGS response (i.e. significant virus accumulation and absence or limited virus symptomatology, reviewed in Burch-Smith et al. 2004). This required the assessment of different cultivars or accessions (Hein et al. 2005), plant growth conditions and inoculation procedure for reliable virus infection. The silencing of endogenous "reporter" genes such as *pds* or *ChlH* (*Magnesium Chelatase* also known as *Sulfur*, Peele et al. 2001), producing visible change in leaf pigmentation as a consequence of their knock-down, provides valuable tools for VIGS vector assessment (Fig. 1a–c).

One of the key milestones in the optimization of RNA viruses as VIGS vectors is the engineering of binary VIGS vectors for Agrobacterium-mediated or biolistic delivery. This method avoids the use of in vitro transcription to generate infectious RNA. This makes virus vectors more compatible for high-throughput silencing screen and a cost-effective approach (Burch-Smith et al. 2004; Meng et al. 2009). In addition, viral genome engineering by using ligation-free cloning approaches facilitates the cloning of large numbers of cDNAs (Liu et al. 2002; Dong et al. 2007).

Enhancement of the silencing response could be achieved for some viruses (TMV, BSMV, Turnip yellow mosaic virus, TYMV) through the expression of direct inverted-repeats folding as hairpin upon transcription (Lacomme et al. 2003; Hein et al. 2005; Pflieger et al. 2008). Hairpin expression is likely to increase the pool of dsRNA and in turn siRNAs (Molnár et al. 2005), suggesting that dsRNA formation is a limiting step in several VIGS systems (Lacomme et al. 2003). Other approaches have been used to optimize VIGS by engineering VIGS vectors with different distribution patterns *in planta*, resulting in an increased invasiveness of meristematic tissues and a more robust silencing response in roots (Valentine et al. 2004). Alternatively, direct agroinfiltration of selected tissues provides an efficient mean to generate a silencing response in a selected area. Direct infiltration of tomato fruit, can produce a visible local silencing in the infiltrated area, bypassing the need for plant infection at an early stage and development of systemic viral infection (Fu et al. 2005).

Demonstration that ectopic expression of a micro(mi)RNA in a modified Cabbage leaf curl geminivirus (CbLCV) generated silencing of selected gene with a comparable efficiency than larger antisense cDNA insert. This approach (so-called "MIR VIGS") has the potential to increase the specificity of the silencing response by engaging the miRNA pathway, therefore offering scope for more precise targeting of gene knock-down (Tang et al. 2010). Recently, it was reported that a TRV silencing vector could trigger RNAi in nematodes; illustrating the potential of VIGS-based approaches in knocking down gene expression of invading pathogen (Valentine et al. 2007; Dubreuil et al. 2009). This potentially opens up a new avenue of research to understand the complex molecular dialog occurring at the host-pathogen interface and importantly for RNA silencing applications in the field of crop protection.

4 Advantages and Limitations of VIGS

The relatively wide host range of some plant virus VIGS vectors makes them suitable for rapidly linking genes to phenotypes in different genetic background. Thus, VIGS offers major advantages to fast track elucidation of molecular bases of agronomical traits in crops species in comparison to transgenic and mutagenesis approaches often restricted to a very limited number of ecotypes or cultivars.

However, plant viruses trigger substantial cellular modifications in their hosts. To fulfill their life cycle, plant viruses use various strategies for the replication of its genome and local movement that include highjacking of the host transcriptional-translational machinery, rearrangement of cellular membranes and of the cytoskeleton and ultimately to access phloem vessels to invade the entire plant (Nelson and Citovsky 2005). During this process, viruses suppress the endogenous RNA silencing response to protect its genome by virus-encoded silencing suppressors. All these events occur concomitantly with the VIGS response and impact on host physiology on top of the phenotype associated with the silenced gene. This requires careful experimental design with the inclusion of suitable controls (such as Mock and control virus-infected plants in addition to VIGS constructs) and biological replicates to deal with plant-to-plant variations associated with the development and extent of the silencing response.

When possible, a careful choice of the cDNA region to be used as a silencing trigger is required in order to minimize (or avoid) off-target silencing effects (i.e. silencing of a distinct gene bearing sufficient sequence identity to the target gene). Several approaches have been developed to deal with this drawback (Cakir et al. 2010), either by (i) choosing a portion of cDNA with little homologies to closely-related genes, (ii) monitoring the expression of the target and potentially "off-target" related genes, (iii) engineering a silencing construct with a different cDNA segment corresponding to the same gene or by (iv) complementation of the silenced phenotype by transient expression of a "cleavage-resistant transgene" (i.e. a similar gene from a different plant species bearing little homologies with the initial target gene (Kerényi et al. 2008). Alternatively, the use of short inverted-repeats (hairpin) offers possibilities for use of short sequence stretches (40–60 nucleotides in length) that can be chosen within highly divergent regions of a gene (Lacomme et al. 2003).

5 Conclusion and Future Prospects

The key advantage of VIGS lies on the potential to rapidly generate gene knock-down, hence mimicking a mutant phenotype in a variety of cultivars and species, where the increasing abundance of microarrays and gene expression profiling studies have generated datasets of differentially expressed genes in different experimental conditions. An increasing number of genome sequencing projects have been undertaken in genetically complex crops with large genomes such as tomato, potato, soybean, barley, maize and wheat. This offers the potential for more finesse in design of gene silencing vectors by defining with greater precision the cDNA region to be used for silencing, allowing for example the knock-down of entire gene families or alternatively specific genes in these species.

The ever increasing knowledge of RNA silencing pathways in different species will have a strong impact in the applications and the continuous improvement of gene silencing (VIGS) systems, enabling us to expand our understanding of the molecular mechanisms underlying key agronomical traits in a large number of crops.

References

Al-Kaff NS, Covey SN, Kreike MM, Page AM, Pinder R, Dale PJ (1998) Transcriptional and posttranscriptional plant gene silencing in response to a pathogen. Science 279:2113–2115

Baulcombe D (2004) RNA silencing in plants. Nature 431:356–363

Brodersen P, Voinnet O (2006) The diversity of RNA silencing pathways in plants. Trends Genet 22:268–280

Bruun-Rasmussen M, Madsen CT, Jessing S, Albrechtsen M (2007) Stability of Barley stripe mosaic virus-induced gene silencing in barley. Mol Plant Microbe Interact 20:1323–1331

Burch-Smith TM, Anderson JC, Martin GB, Dinesh-Kumar SP (2004) Applications and advantages of virus-induced gene silencing for gene function studies in plants. Plant J 39:734–746

Cakir C, Gillespie ME, Scofield SR (2010) Rapid determination of gene function by virus-induced gene silencing in wheat and barley. Crop Sci 50:S77–S84

Constantin GD, Grønlund M, Johansen IE, Stougaard J, Lund OS (2008) Virus-induced gene silencing (VIGS) as a reverse genetic tool to study development of symbiotic root nodules. Mol Plant Microbe Interact 21:720–727

Ding XS, Schneider WL, Chaluvadi SR, Mian MA, Nelson RS (2006) Characterization of a Brome mosaic virus strain and its use as a vector for gene silencing in monocotyledonous hosts. Mol Plant Microbe Interact 19:1229–1239

Donaire L, Barajas D, Martínez-García B, Martínez-Priego L, Pagán I, Llave C (2008) Structural and genetic requirements for the biogenesis of tobacco rattle virus-derived small interfering RNAs. J Virol 82:5167–5177

Dong Y, Burch-Smith TM, Liu Y, Mamillapalli P, Dinesh-Kumar SP (2007) A ligation-independent cloning tobacco rattle virus vector for high-throughput virus-induced gene silencing identifies roles for NbMADS4-1 and -2 in floral development. Plant Physiol 145:1161–1170

Dunoyer P, Voinnet O (2005) The complex interplay between plant viruses and host RNA-silencing pathways. Curr Opin Plant Biol 8:415–423

Dubreuil G, Magliano M, Dubrana MP, Lozano J, Lecomte P, Favery B, Abad P, Rosso MN (2009) Tobacco rattle virus mediates gene silencing in a plant parasitic root-knot nematode. J Exp Bot 60:4041–4050

Faivre-Rampant O, Gilroy EM, Hrubikova K, Hein I, Millam S, Loake GJ, Birch P, Taylor M, Lacomme C (2004) Potato virus X-induced gene silencing in leaves and tubers of potato. Plant Physiol 134:1308–1316

Fitzmaurice WP, Holzberg S, Lindbo JA, Padgett HS, Palmer KE, Wolfe GM, Pogue GP (2002) Epigenetic modification of plants with systemic RNA viruses. OMICS 6:137–151

Fofana IB, Sangaré A, Collier R, Taylor C, Fauquet CM (2004) A geminivirus-induced gene silencing system for gene function validation in cassava. Plant Mol Biol 56:613–624

Fu DQ, Zhu BZ, Zhu HL, Jiang WB, Luo YB (2005) Virus-induced gene silencing in tomato fruit. Plant J 43:299–308

Fulci V, Macino G (2007) Quelling: post-transcriptional gene silencing guided by small RNAs in Neurospora crassa. Curr Opin Microbiol 10:199–203

Gilroy EM, Hein I, van der Hoorn R, Boevink PC, Venter E, McLellan H, Kaffarnik F, Hrubikova K, Shaw J, Holeva M, López EC, Borras-Hidalgo O, Pritchard L, Loake GJ, Lacomme C, Birch PR (2007) Involvement of cathepsin B in the plant disease resistance hypersensitive response. Plant J 52:1–13

Goodin MM, Zaitlin D, Naidu RA, Lommel SA (2008) *Nicotiana benthamiana*: its history and future as a model for plant-pathogen interactions. Mol Plant-Microbe Interact 21:1015–1026

Gosselé V, Faché I, Meulewaeter F, Cornelissen M, Metzlaff M (2002) SVISS—a novel transient gene silencing system for gene function discovery and validation in tobacco plants. Plant J 32:859–866

Grishok A (2005) RNAi mechanisms in caenorhabditis elegans. FEBS Lett 579:5932–5939

Hamilton AJ, Baulcombe DC (1999) A species of small antisense RNA in posttranscriptional gene silencing in plants. Science 286:950–952

Hein I, Barciszewska-Pacak M, Hrubikova K, Williamson S, Dinesen M, Soenderby IE, Sundar S, Jarmolowski A, Shirasu K, Lacomme C (2005) Virus-induced gene silencing-based functional characterization of genes associated with powdery mildew resistance in barley. Plant Physiol 138:2155–2164

Herr AJ, Molnàr A, Jones A, Baulcombe DC (2006) Defective RNA processing enhances RNA silencing and influences flowering of arabidopsis. Proc Natl Acad Sci U S A 103:14994–15001

Himber C, Dunoyer P, Moissiard G, Ritzenthaler C, Voinnet O (2003) Transitivity-dependent and -independent cell-to-cell movement of RNA silencing. EMBO J 22:4523–4533

Hiriart JB, Aro EM, Lehto K (2003) Dynamics of the VIGS-mediated chimeric silencing of the Nicotiana benthamiana ChlH gene and of the tobacco mosaic virus vector. Mol Plant Microbe Interact 16:99–106

Ho T, Pallett D, Rusholme R, Dalmay T, Wang H (2006) A simplified method for cloning of short interfering RNAs from Brassica juncea infected with Turnip mosaic potyvirus and Turnip crinkle carmovirus. J Virol Methods 136:217–223

Ho T, Wang H, Pallett D, Dalmay T (2007) Evidence for targeting common siRNA hotspots and GC preference by plant dicer-like proteins. FEBS Lett 581:3267–3272

Holzberg S, Brosio P, Gross C, Pogue GP (2002) Barley stripe mosaic virus-induced gene silencing in a monocot plant. Plant J 30:315–327

Igarashi A, Yamagata K, Sugai T, Takahashi Y, Sugawara E, Tamura A, Yaegashi H, Yamagishi N, Takahashi T, Isogai M, Takahashi H, Yoshikawa N (2009) Apple latent spherical virus vectors for reliable and effective virus-induced gene silencing among a broad range of plants including tobacco, tomato, Arabidopsis thaliana, cucurbits, and legumes. Virology 386:407–416

Karrer EE, Beachy RN, Holt CA (1998) Cloning of tobacco genes that elicit the hypersensitive response. Plant Mol Biol 36:681–690

Katiyar-Agarwal S, Morgan R, Dahlbeck D, Borsani O, Villegas A Jr, Zhu JK, Staskawicz BJ, Jin H (2006) A pathogen-inducible endogenous siRNA in plant immunity. Proc Natl Acad Sci U S A 103:18002–18007

Kerényi Z, Mérai Z, Hiripi L, Benkovics A, Gyula P, Lacomme C, Barta E, Nagy F, Silhavy D (2008) Inter-kingdom conservation of mechanism of nonsense-mediated mRNA decay. EMBO J 27:1585–1595

Kjemtrup S, Sampson KS, Peele CG, Nguyen LV, Conkling MA, Thompson WF, Robertson D (1998) Gene silencing from plant DNA carried by a geminivirus. Plant J 14:91–100

Kumagai MH, Donson J, della-Cioppa G, Harvey D, Hanley K, Grill LK (1995) Cytoplasmic inhibition of carotenoid biosynthesis with virus-derived RNA. Proc Natl Acad Sci U S A 92:1679–1683

Lacomme C, Hrubikova K, Hein I (2003) Enhancement of virus-induced gene silencing through viral-based production of inverted-repeats. Plant J 34:543–553

Lacomme C, Pogue GP, Wilson TMA, Santa Cruz S (2001) Plant viruses. In: Ring CJA, Blair E (eds) Genetically engineered viruses: development and applications. BIOS Scientific Publishing Limited, Oxford

Lacomme C, Santa Cruz S (1999) Bax-induced cell death in tobacco is similar to the hypersensitive response. Proc Natl Acad Sci U S A 96:7956–7961

Lindbo JA, Dougherty WG (1992) Pathogen-derived resistance to a potyvirus: immune and resistant phenotypes in transgenic tobacco expressing altered forms of a potyvirus coat protein nucleotide sequence. Mol Plant Microbe Interact 5:144–153

Lindbo JA, Silva-Rosales L, Proebsting WM, Dougherty WG (1993) Induction of a highly Specific antiviral state in transgenic plants: implications for regulation of gene expression and virus resistance. Plant Cell 5:1749–1759

Liu Y, Schiff M, Dinesh-Kumar SP (2002) Virus-induced gene silencing in tomato. Plant J 31:777–786

Lu R, Malcuit I, Moffett P, Ruiz MT, Peart J, Wu AJ, Rathjen JP, Bendahmane A, Day L, Baulcombe DC (2003) High throughput virus-induced gene silencing implicates heat shock protein 90 in plant disease resistance. EMBO J 22:5690–5699

Matthews REF (1991) Plant virology, 3rd edn. Academic Press Inc, London

Meng Y, Moscou MJ, Wise RP (2009) *Blufensin1* negatively impacts basal defense in response to barley powdery mildew. Plant Physiol 149:271–285

Mlotshwa S, Pruss GJ, Vance V (2008) Small RNAs in viral infection and host defense. Trends Plant Sci 13:375–382

Molnár A, Csorba T, Lakatos L, Várallyay E, Lacomme C, Burgyán J (2005) Plant virus-derived small interfering RNAs originate predominantly from highly structured single-stranded viral RNAs. J Virol 79:7812–7818

Moore CJ, Sutherland PW, Forster RL, Gardner RC, MacDiarmid RM (2001) Dark green islands in plant virus infection are the result of posttranscriptional gene silencing. Mol Plant Microbe Interact 14:939–946

Navarro L, Dunoyer P, Jay F, Arnold B, Dharmasiri N, Estelle M, Voinnet O, Jones JD (2006) A plant miRNA contributes to antibacterial resistance by repressing auxin signaling. Science 312:436–439

Nelson RS, Citovsky V (2005) Plant viruses. Invaders of cells and pirates of cellular pathways. Plant Physiol 138:1809–1814

Pantaleo V, Szittya G, Burgyán J (2007) Molecular bases of viral RNA targeting by viral small interfering RNA-programmed RISC. J Virol 81:3797–3806

Peele C, Jordan CV, Muangsan N, Turnage M, Egelkrout E, Eagle P, Hanley-Bowdoin L, Robertson D (2001) Silencing of a meristematic gene using geminivirus-derived vectors. Plant J 27:357–366

Pflieger S, Blanchet S, Camborde L, Drugeon G, Rousseau A, Noizet M, Planchais S, Jupin I (2008) Efficient virus-induced gene silencing in Arabidopsis using a 'one-step' TYMV-derived vector. Plant J 56:678–690

Ratcliff F, Harrison BD, Baulcombe DC (1997) A similarity between viral defense and gene silencing in plants. Science 276:1558–1560

Robertson D (2004) VIGS vectors for gene silencing: many targets, many tools. Annu Rev Plant Biol 55:495–519

Rommens CM, Salmeron JM, Baulcombe DC, Staskawicz BJ (1995) Use of a gene expression system based on potato virus X to rapidly identify and characterize a tomato Pto homolog that controls fenthion sensitivity. Plant Cell 7:249–257

Schwach F, Vaistij FE, Jones L, Baulcombe DC (2005) An RNA-dependent RNA polymerase prevents meristem invasion by potato virus X and is required for the activity but not the production of a systemic silencing signal. Plant Physiol 138:1842–1852

Scofield SR, Huang L, Brandt AS, Gill BS (2005) Development of a virus-induced gene-silencing system for hexaploid wheat and its use in functional analysis of the Lr21-mediated leaf rust resistance pathway. Plant Physiol 138:2165–2173

Szittya G, Silhavy D, Molnár A, Havelda Z, Lovas A, Lakatos L, Bánfalvi Z, Burgyán J (2003) Low temperature inhibits RNA silencing-mediated defence by the control of siRNA generation. EMBO J 22:633–640

Takken FL, Luderer R, Gabriels SH, Westerink N, Lu R, de Wit PJ, Joosten MH (2000) A functional cloning strategy, based on a binary PVX-expression vector, to isolate HR-inducing cDNAs of plant pathogens. Plant J 24:275–283

Tang Y, Wang F, Zhao J, Xie K, Hong Y, Liu Y (2010) Virus-based microRNA expression for gene functional analysis in plants. Plant Physiol 153:632–641

Thomas CL, Jones L, Baulcombe DC, Maule AJ (2001) Size constraints for targeting post-transcriptional gene silencing and for RNA-directed methylation in Nicotiana benthamiana using a potato virus X vector. Plant J 25:417–425

Tuttle JR, Idris AM, Brown JK, Haigler CH, Robertson D (2008) Geminivirus-mediated gene silencing from cotton leaf crumple virus is enhanced by low temperature in cotton. Plant Physiol 148:41–50

Turnage MA, Muangsan N, Peele CG, Robertson D (2002) Geminivirus-based vectors for gene silencing in arabidopsis. Plant J 30:107–114

Valentine T, Shaw J, Blok VC, Phillips MS, Oparka KJ, Lacomme C (2004) Efficient virus-induced gene silencing in roots using a modified tobacco rattle virus vector. Plant Physiol 136:3999–4009

Valentine TA, Randall E, Wypijewski K, Chapman S, Jones J, Oparka KJ (2007) Delivery of macromolecules to plant parasitic nematodes using a tobacco rattle virus vector. Plant Biotechnol J 5:827–834

van Kammen A (1997) Virus-induced gene silencing in infected and transgenic plants. Trends Plant Sci 2:409–411

Voinnet O (2005) Non-cell autonomous RNA silencing. FEBS Lett 579:5858–5871

Voinnet O, Pinto YM, Baulcombe DC (1999) Suppression of gene silencing: a general strategy used by diverse DNA and RNA viruses of plants. Proc Natl Acad Sci U S A 96:14147–14152

Wege S, Scholz A, Gleissberg S, Becker A (2007) Highly efficient virus-induced gene silencing (VIGS) in California poppy (Eschscholzia Californica): an evaluation of VIGS as a strategy to obtain functional data from non-model plants. Ann Bot 100:641–649

Wypijewski K, Hornyik C, Shaw JA, Stephens J, Goraczniak R, Gunderson SI, Lacomme C (2009) Ectopic 5′ splice sites inhibit gene expression by engaging RNA surveillance and silencing pathways in plants. Plant Physiol 151:955–965

Yang SJ, Carter SA, Cole AB, Cheng NH, Nelson RS (2004) A natural variant of a host RNA-dependent RNA polymerase is associated with increased susceptibility to viruses by nicotiana benthamiana. Proc Natl Acad Sci U S A 101:6297–6302

Emerging Antibody-based Products

Kevin J. Whaley, Josh Morton, Steve Hume, Ernie Hiatt, Barry Bratcher, Victor Klimyuk, Andrew Hiatt, Michael Pauly and Larry Zeitlin

Abstract Antibody-based products are not widely available to address many global health challenges due to high costs, limited manufacturing capacity, and long manufacturing lead times. There are now tremendous opportunities to address these industrialization challenges as a result of revolutionary advances in plant virus-based transient expression. This review focuses on some antibody-based products that are in preclinical and clinical development, and have scaled up

K. J. Whaley (✉) · A. Hiatt · M. Pauly · L. Zeitlin
Mapp Biopharmaceutical Inc, 6160 Lusk Blvd, Suite C105, San Diego, CA 92121, USA
e-mail: Kevin.whaley@mappbio.com

A. Hiatt
e-mail: Andy.hiatt@mappbio.com

M. Pauly
e-mail: Michael.pauly@mappbio.com

L. Zeitlin
e-mail: Larry.zeitlin@mappbio.com

J. Morton · S. Hume · E. Hiatt · B. Bratcher
Kentucky BioProcessing, LLC 3700 Airpark Drive, Owensboro, KY 42301, USA
e-mail: JMorton@kbpllc.com

S. Hume
e-mail: SHume@kbpllc.com

E. Hiatt
e-mail: EHiatt@kbpllc.com

B. Bratcher
e-mail: BBratcher@kbpllc.com

V. Klimyuk
Icon Genetics GmbH, Biozentrum Halle, Weinbergweg 22, Halle (Saale), D-06120, Germany
e-mail: Klimyuk@icongenetics.de

Current Topics in Microbiology and Immunology (2012) 375: 107–126
DOI: 10.1007/82_2012_240

Published Online: 8 July 2012

manufacturing and purification (mg of purified mAb/kg of biomass). Plant virus-based antibody products provide lower upfront cost, shorter time to clinical and market supply, and lower cost of goods (COGs). Further, some plant virus-based mAbs may provide improvements in pharmacokinetics, safety and efficacy.

List of abbreviations and acronyms

mAb	monoclonal antibody
HIV	human immunodeficiency virus
HSV	herpes simplex virus
RSV	respiratory syncytial virus
ΔXF	xylose and fucose knockout
CHO	Chinese hamster ovary
gD	glycoprotein D
NHL	Non-Hodgkin's lymphoma
F-L	follicular B-cell lymphoma
GnGn	human like biantennary N-glycans with terminal N-acetylglucosamine on each branch
CHT	ceramic hydroxyapatite

Contents

1 Introduction

Monoclonal antibodies (mAbs) have been therapeutically and financially successful in developed markets, but global access is limited due in part to industrialization challenges. Now, rapid and versatile manufacturing platform technologies based on plant virus-based transient expression allow fast and inexpensive manufacturing of mAbs for large, cost-sensitive markets. This chapter reviews emerging antibody-based products that could be competitive and commercially viable when the attributes of plant virus-based transient expression are utilized in manufacturing. Rather than a review of all mAbs that have been expressed with transient viral systems, the focus of this review is on some mAbs that have begun the transition to clinical development (i.e., animal efficacy data and scaled up manufacturing).

2 The Critical Path for Antibody-based Products

The Critical Path process (FDA 2004) starts with the selection of a lead candidate for preclinical and clinical development. Developers of antibody-based products must negotiate three crucial scientific/technical dimensions (industrialization, safety, efficacy) on the critical path from scientific innovation to commercial product (Table 1). The three science dimensions are interdependent; the vast majority of development costs are attributable to these dimensions. Business dimensions (e.g., obtaining capital, intellectual property, marketing, distribution) are outside the scope of this review.

The main causes of failure in clinical trials include safety problems and lack of effectiveness, but problems in industrialization (physical design, characterization, manufacturing scale-up and quality control) routinely derail or delay development programs. The challenges involved in successful industrialization are complex; unfortunately, the challenges are highly underrated in the broader scientific community.

Table 1 **Critical path and dimensions (adapted from FDA 2004)**

Dimensions	*Critical Path*			
	Lead mAb Selection	**Preclinical Development**	**Clinical Development**	**Regulatory Approval**
Industrialization	Establish scalable manufacturing and purification documentation manufacturing	Manufacturing and purification scale-up Commercial	Refine specifications and	
Safety	In vitro	Animal models	Human safety trials	Post-market safety follow up
Efficacy	In vitro	Animal models	Human efficacy trials	Post-market efficacy follow up

3 Industrialization of Antibody-based Products using Plant Viruses

Plant virus-based transient infection of *Nicotiana* has accelerated the industrialization component of the critical path for mAbs (Table 1), and consequently the preclinical and clinical development of antibody-based products produced in plants. These transient expression systems, especially magnICON (Giritch et al. 2006), Gemini virus (Huang et al. 2010), and Cowpea mosaic virus (Sainsbury and Lomonossoff, 2008; Sainsbury et al. 2010) have been shown to produce full length antibodies. Until recently, most production of mAbs in plants had been at small scale (see review; DeMuynck et al. 2010). Now, a number of institutions are manufacturing mAbs in *Nicotiana* for preclinical and clinical studies under good manufacturing practices (GMP). GMP manufacturing of mAbs with viral based transient expression is currently conducted by Kentucky BioProcessing (KBP; Owensboro, KY) and Icon Genetics (Halle, Germany) utilizing magnICON technology, while the Fraunhofer Institute (Aachen, Germany) is using CPMV.

The first plant virus system to be adequately industrialized for the production of full length antibodies "on demand" in large quantities, at low cost is magnICON (Giritch et al. 2006; Pogue et al. 2010; Klimyuk et al. 2012). The technology is in essence an *en masse* infiltration of whole mature plants with a highly dilute *Agrobacterium* suspension carrying T-DNAs encoding viral replicons. The result is a high copy number of RNA molecules that encode the desired antibody, reducing the time needed to produce the mAb to 6–8 days. Moreover, increasing the volume of mAb–containing biomass does not require drastic changes in growing conditions, infection procedures, and is directly scalable. No genes are incorporated into the plant genome, and as a result, there is no risk of propagation of the transgene from pollen, seeds, or other routes. Further, no intact and replication-competent virus is produced, eliminating the risk of virus-mediated spreading of the recombinant genes. Finally, the entire production system can be performed indoors in enclosed growth rooms, providing an additional layer of environmental security and quality control.

Since wild-type *N. benthamiana* glycosylates proteins differently than mammalian expression systems (Gomord et al. 2010), the magnICON expression technology has been successfully coupled with advantageous glycosylation technology by Icon (Halle, Germany). *N. benthamiana*, like other plants, produces the same core glycan as found in mammals, but uses xylose (which generally is not found in mammals) and fucose in a non-mammalian linkage (alpha 1,3). Because of the potential for the novel plant glycans to affect pharmacokinetics as well as immunogenicity in humans, it is highly desirable to produce mAbs in plants that have been modified to generate mammalian-like glycans. Now, with the development of transgenic strains (ΔXF) of *N. benthamiana* with fucosyl– and xylosyl– transferase knocked down by RNAi (Schahs et al. 2007; Strasser et al. 2008, 2009; Castilho et al. 2011), plants can produce mAbs with glycoforms that are essentially mammalian (Vezina et al. 2009). The resulting glycoforms on mAbs produced in these plants are more homogeneous than many FDA-approved mAbs produced in mammalian cell culture (Whaley et al. 2011; Zeitlin et al. 2011); obtaining a consistent glycoform profile in production may be desirable from a quality and regulatory perspective. Of particular note for the development of mAbs where antibody dependent cellular cytotoxicity (ADCC) is an important mechanism of action (e.g., anti-cancer antigen mAbs), the predominant glycoform (human like biantennary N-glycans with acetylglucosamine on each branch, GnGn) is one that is ideal for ADCC activity—absence of core fucose has been shown to increase ADCC activity dramatically. For example, de-fucosylated, glycosylated Herceptin (trastuzumab) was shown to be at least 50-fold more active in the efficacy of Fc-gamma receptor IIIa (Fc RIIIa)-mediated ADCC than herceptin with alpha-1,6-linked fucose residues (Shields et al. 2002). Similar results were reported for Rituximab and other mAbs (Shinkawa et al. 2003; Niwa et al. 2004). Unfortunately, most recombinant mAbs are currently produced in a fashion such that the antibody protein is present as a mixture of glycans, in which the more active glycoform (e.g., de-fucosylated) may be present only in minor amounts or as a component of five or more glycans. All currently marketed mAbs are only available as complex, heterogeneous glycoforms as a result of their mammalian cell culture origin. More homogeneous glycoforms could very well reduce the dosages and improve efficacy of therapeutically beneficial recombinant mAbs.

Downstream processing of mAbs is complex and costly (accounting for as much as 50–80% of total manufacturing costs). Currently, downstream purification of mAbs produced in virus-based transient plant systems is dependent on standard antibody purification methods like protein A affinity capture (Klimyuk et al. 2012). In the future, use of a plant virus-derived immunoabsorbent (based on a tobamovirus displaying protein A) may be an especially attractive tool for the purification of *Nicotiana*-made mAbs (Werner et al. 2006). Kentucky BioProcessing is investigating cation exchange chromatography, in conjunction with precipitation for removal of host cell proteins, as an alternative means of mAb capture.

4 Idiotype Immunoglobulins for Non-Hodgkin's Lymphoma

Non-Hodgkin's lymphoma (NHL) is the most common hematologic malignancy in the United States, with over 60,000 estimated new cases this year. A quarter of these cases are follicular B cell lymphoma (FL), a disease still considered incurable; most patients subjected to standard treatments relapse (Bendandi, 2008). Although B cell tumors in general and FL in particular are characterized by extreme variability in treatment and prognosis, they share certain features that make them ideal for the development of patient-specific cancer vaccines. Each clone of malignant B cells expresses a unique cell surface immunoglobulin (Ig), which serves as a tumor-specific marker. Therefore, newer approaches have focused on active immunotherapy through vaccination with the patient's own idiotype (Id) conjugated to a strong immunogen such as keyhole limpet hemocyanin (KLH) and administered with an adjuvant-like granulocyte–macrophage colony stimulating factor (GM-CSF). The history of Id vaccination for follicular lymphoma is >20 years old, and numerous studies have demonstrated that upon vaccination during chemotherapy-induced remission, a large percentage of patients develop an antigen-specific immune response and the responders have a superior clinical outcome (Bendandi, 2006). However, all three phase III clinical trials aiming at Id vaccine regulatory approval for follicular lymphoma have failed to achieve their main clinical end points mainly because of serious pitfalls in their study design or in their conduct (Bendandi, 2009; Lopez-Diaz and Inoges 2009). Several methods for immunoglobulin-based vaccine production have been used, including development of hybridomas that secrete patient's tumor-specific Ig, cloning of Ig variable regions and expression of full Ig molecules in *Escherichia coli*, insect or mammalian cells or expression of cloned variable regions as single-chain variable fragments (scFv) in plants using plant viral vectors (McCormick et al. 2003, 2008). Absence of efficacy in a phase 3 trial with scFvs (Baculovirus-derived) may indicate that a whole Ig approach is needed.

4.1 Industrialization

Individualized vaccines are highly dependent on rapid manufacturing capability. The cost, production time and availability of tailor-made, patient- and tumor-specific Id antibodies have been significant hurdles to previous developments in this field. Protein A affinity capture purification and anion exchange yielded idiotype immunoglobulins (magnICON expressed) of pharmaceutical quality that ranged from 10–200 mg purified IgG/ kg biomass (Fig. 1).

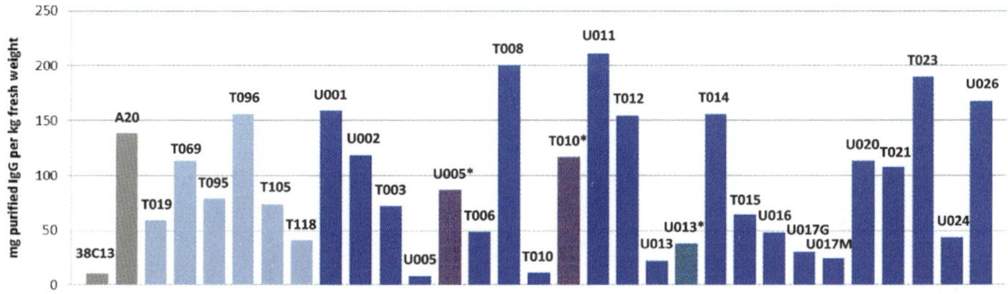

Fig. 1 Idiotype Immunoglobulin yield (mg purified/kg biomass). The data includes idiotypes for animal models (38C13 and A20) as well as 26 different idiotypes for preclinical (T019; T069; T095; T096; T105 and T118) and clinical (remaining) studies. * - yield using optimized downstream process

4.2 Safety and Efficacy

Testing of magnICON–expressed Id antigens in two animal models (38C13 and A20) showed increases in survival equivalent to the gold standard hybridoma-derived IgM-KLH and both Ids provided significant protection compared to controls (Bendandi et al. 2010). Initial data from a phase 1 safety trial on 12 patients addressing the safety of full idiotype mAbs have been reported (Bendandi et al. 2011a). Patients enrolled in a recent study, entered into the vaccination stage after maintaining at least four month complete (CR) or partial response (PR) following bendamustine treatment (Bendandi, 2011b). Each patient was subjected to 12 doses of anti-Id mAb with a 4 week interval between the vaccinations. Each dose included a conjugate of 0.5 mg of id mAb and 0.5 mg of keyhole limpet hemocyanin. The dose was co-administered with 0.125 mg of GM-CSF. No patient has yet completed the entire vaccination protocol and only one patient has relapsed/progressed while receiving the id mAb vaccine. The intermediate results of vaccine tolerability by patients demonstrate that mAb–based vaccine is well tolerated in patients with relapsed follicular lymphoma following bendamustine–based chemotherapy.

5 Ebola Antibody-based Immunoprotectant

No countermeasures currently exist for the prevention or treatment of the severe sequelae of filovirus (i.e., ebola, marburg) infections. To overcome this limitation in biodefense and public health preparedness, monoclonal antibodies (mAbs) were generated which could be used in humans as immunoprotectants for ebola (EBOV) (Zeitlin et al. 2011). Monoclonal antibodies against marburg and other hemorrhagic fever viruses are at the lead selection stage; multipurpose products that are effective against a variety of pathogens are a priority for BARDA (see Strategic Plan, 2011).

5.1 Industrialization

Three murine mAbs (6D8, 13C6, 13F6) against the ebola virus glycoprotein, shown to protect against lethal ebola challenge in mice (Wilson et al. 2000), were subjected to deimmunisation. Specifically, point mutations were introduced into the variable regions to remove predicted human T cell epitopes, and the variable regions joined to human constant regions to generate mAbs (h-6D8, h-13C6, h-13F6) appropriate for development for human use. Manufacturing runs of 500 kg scale–using magnICON and ΔXF *N. benthamiana* (Strasser et al. 2008) were performed and mAb purified by protein A and ceramic hydroxyapatite chromatography (Zeitlin et al. 2011). The purity and contaminant levels were appropriate for human injection.

5.2 Efficacy

The efficacy of all three ebola mAbs has been shown in the mouse model (Zeitlin et al. 2011). Interestingly, the glycosylation patterns (majority GnGn) conferred by the ΔXF *N. benthamiana* resulted in improved efficacy compared to the identical mAb expressed in mammalian NS0 cells. Binding studies using Fcγ receptors revealed enhanced binding of non-fucosylated mAb to mouse and human FcγRIII, a receptor that mediates ADCC activity. Together, the results indicate that the presence of Fc N-glycans enhances the protective efficacy of the Ebola mAbs, and that mAbs manufactured with uniform glycosylation and a higher potency glycoform offer promise as biodefense therapeutics. These mAbs are currently being tested in non-human primates for post-exposure efficacy against lethal challenge.

6 Antibody-based Immune Complexes Ebola Vaccine

No vaccine against ebola currently exists, but a number of strong candidates have emerged. While some have yielded good efficacy in animal models, they have practical shortcomings. Most existing vaccine candidates are genetically modified live viruses; vaccines of this sort require very careful conditions of storage and have a tendency to lose potency over a period of months. Based on a previously described recombinant immune complex vaccine strategy (Chargelegue et al. 2005), an ebola vaccine candidate was designed in which the viral surface glycoprotein is manufactured as a fusion to a monoclonal antibody (h-6D8) that recognizes an epitope in the glycoprotein, resulting in Nicotiana production of ebola immune complexes (EICs).

6.1 Industrialization

The antibody-based vaccine was designed by fusing h-6D8 with ebola virus GP1 and expressed using the Gemini virus system (Huang et al. 2009; Phoolcharoen et al. 2011a). The resulting molecules' opposite ends attract each other, like a group of rodshaped magnets. When the vaccine molecules bind to each other, they form an aggregate (or repeating array) called an ebola immune complex (EIC). EIC were purified via protein G affinity chromatography at 2–3 mg/kg; purity was ~90 %.

6.2 Efficacy

The immunogenicity (Phoolcharoen et al. 2011a) and efficacy (Phoolcharoen et al. 2011b) of the EBOV EIC vaccine candidate has been demonstrated in mice. Mice in each of the vaccination regimes lost weight over the course of the 14 day observation period, indicating none of the immunized animals had sterilizing immunity. Four doses of EIC (10 μg) co-delivered with Toll-like receptor agonist polyinosinic:polycytidylic acid (PIC) or co-delivered with PIC and alum protected 80 % of mice against lethal Ebola challenge (similar to GP-VRP vaccinated positive control mice). Mice vaccinated with four doses of EIC co-delivered with PIC or h-EIC co-delivered with PIC and alum had significantly higher anti-EBOV IgG production compared with h-EIC alone. These two vaccination regimens resulted in the highest level of neutralizing antibody titer.

7 Respiratory Syncytial Virus mAb Immunoprophylaxis

Although an RSV mAb is commercially available (Synagis®, MedImmune), and is effective for RSV immunoprophylaxis in at-risk neonates, ongoing analyses on the pharmacoeconomics demonstrate that this mAb is not cost-effective in actual use, nor even when its use is restricted very tightly to its labeled population (Joffe et al. 1999; Vogel et al. 2002; Roeckl-Wiedmann et al. 2003). Due in part to cost, *a majority of the world's population does not have access to an effective RSV mAb; in fact, it is unavailable in 7 of the 10 most populous countries, including India and China*. If the costs of RSV immunoprophylaxis were lowered, the patient populations in which the drug would be cost-effective would be dramatically expanded. For example, 5–10 % of elderly patients in long-term care facilities develop RSV infections per year with rates of pneumonia and death of 10–20 % and 2–5 % respectively (Han et al. 1999; Falsey and Walsh, 2000). Based on manufacturing costs at KBP's existing facility, and assuming 100 mg purified RSV mAb/kg biomass, costs of goods for Nicotiana-derived mAbs are approximately one twentieth the wholesale cost of Synagis.

7.1 Industrialization

MB-RSV was generated using magnICON and ΔXF *N. benthamiana*. MB-RSV was purified by protein A chromatography, yielding 180 mg purified mAb/kg biomass (Zeitlin et al. 2012).

7.2 Efficacy

Cotton rats received mAb via the intramuscular (IM) route one day prior to intranasal challenge. Four days post-challenge, animals were euthanized and lung tissue harvested to determine viral titer. MB-RSV displayed an expected dose-dependent reduction in viral titer of 2–3 orders of magnitude identical to that of Synagis® (Zeitlin et al. 2012).

8 Antibody for Breast Cancer Therapy

Human epidermal growth factor receptor 2 (HER2/neu) is an oncogene involved in abnormal cell growth in breast cancer and is a target for the humanized monoclonal antibody (mAb) trastuzumab (Herceptin®) (Hudziak et al. 1989), which was approved by the U.S. food and drug administration for the treatment of HER2/neu-overexpressing breast tumours. HER2/neu is overexpressed in 20–30 % of metastatic breast cancer patients where its overexpression results in the disruption of normal signaling pathways, causing the loss of cell growth regulation and the development of resistance to apoptosis. Trastuzumab induces antibody-dependent cellular cytotoxicity (ADCC), inhibits HER2-mediated signaling, and prevents cleavage of the extracellular domain of HER2 (Hudis 2007). In HER2- positive breast cancer, trastuzumab has shown a survival advantage in early and metastatic disease and is now the standard of care (Piccaart-Gebhart et al. 2005; Slamon et al. 2001; Smith et al. 2007). If long-term (years) cancer therapy with mAbs becomes routine, the pharmacoeconomics and supply requirements may change substantially. For example, Herceptin (Genentech) can cost a breast cancer patient more than $37,000 for a year of treatment.

8.1 Industrialization

Plant made Trastuzumab (PMT) was expressed with 35S-based constructs as well as the magnICON viral vectors (Komarova et al. 2011); expression using magnICON was approximately twice that of the 35S–based vectors. The yield of PMT expressed from magnICON was between 200 and 300 mg/g FW depending on the experiment.

8.2 *Efficacy*

PMT demonstrated similar in vitro growth inhibition of a breast cancer cell line. When tested in a xenograft mouse model of human Her+ 2 ovarian cancer, PMT was significantly more effective than trastuzumab produced in mammalian cells in suppressing tumor volume (Komarova et al. 2011). The cause of this improved in vivo activity compared to mammalian trastuzumab is unclear. Since equivalent binding activity was demonstrated between PMT and trastuzumab, it is likely the differences in efficacy are related to glycosylation mediated effects on binding to Fc receptors.

9 Ricin Antibodies

Ricin is a member of the type II ribosome-inactivating protein (RIP) family of toxins that includes abrin, Shiga toxin from *Shigella dysenteriae*, and Shiga-like toxins from *Escherichia coli* (Olsnes 2004; Stirpe 2004). Because the toxin can be lethal to humans following injection, inhalation and possibly ingestion, ricin has been classified by the centers for disease control and prevention considered a category B biothreat agent.

The development of an antibody–based therapeutic for treatment of individuals following exposure to ricin via the systemic or mucosal routes is important. A collection of ricin-neutralizing monoclonal antibodies (mAbs) against both RTA and RTB (Neal et al. 2010; O'Hara et al. 2010; Yermakova and Mantis 2011) have been produced and characterized.

9.1 *Industrialization*

GD12 is a murine monoclonal IgG_1 (mAb) that recognizes an immunodominant linear neutralizing epitope (163-TLARSFIICIQM-174) on the A subunit (RTA) of ricin toxin (O'Hara et al. 2012). With the long-term goal of using GD12 as a potential countermeasure against ricin intoxication, a chimeric derivative of GD12 (cGD12) has been produced in which the murine heavy and light chain variable regions were fused to a human IgG_1 framework. The chimeric mAb utilized magnICON and ΔXF *N. benthamiana* and purified by protein A chromatography.

9.2 Efficacy

Plant expressed cGD12 retained epitope specificity and ricin neutralizing activity similar to the parental murine mAb (O'Hara et al. 2012). Passive administration of cGD12 (10 μg; 0.5 mg/kg) to mice by intraperitoneal injection protected the animals against a systemic 10 LD_{50} ricin challenge. In a post-exposure setting, the murine and chimeric mAbs administered as much as 6 h after toxin challenge were each capable of rescuing mice from toxin-induced death, revealing the potential of GD12 to serve as both a prophylactic and therapeutic for ricin intoxication (O'Hara et al. 2012).

10 Antibody for Treating Fungal Infections

There is an increasing interest in the development of therapeutic antibodies (Ab) to improve the control of fungal pathogens, but none of these reagents are available for clinical use. A murine monoclonal antibody (mAb 2G8) targeting ß-glucan (Capodicasa et al. 2011), a cell wall polysaccharide common to most pathogenic fungi, has conferred significant protection against *Candida albicans*, *Aspergillus fumigatus* and *Cryptococcus neoformans* in animal models.

10.1 Industrialization

Chimeric mouse–human Ab derivatives from mAb 2G8 were transiently expressed (CAMV/TMV) in *Nicotiana benthamiana* plants and purified from leaves with high yields (approximately 50 mg mAb/kg of plant tissues) (Capodicasa et al. 2011). The N-glycan structures found on the heavy chain were exclusively those of the typical plant complex type.

10.2 Efficacy

Plant-derived mAbs fully retained the ß-glucan-binding specificity and the anti-fungal activities of the cognate murine mAb against *C. albicans*. The mAbs recognized preferentially ß1,3- linked glucan molecules present at the fungal cell surface and directly inhibited the growth of *C. albicans* and its adhesion to human epithelial cells in vitro. In addition, the mAb promoted *C. albicans* killing by isolated, human polymorphonuclear neutrophils in ex vivo assays, and conferred significant antifungal protection in both mucosal and systemic infection models of candidiasis.

In the systemic infection model, a single dose of each mAb was administered to mice followed by an intravenous challenge with *C. albicans*, and protection was evaluated by measuring fungal invasion of kidneys at day 2 postinfection (Capodicasa et al. 2011). In these experiments, treatment with the chimeric IgG resulted in a significant reduction in fungal load in kidneys when compared to controls. mAbs were also tested for protection in a rat model of experimental vaginitis in which animals received a 'therapeutic' mAb treatment at 1, 24 and 48 h postintravaginal infection with *C. albicans*; the protective effect was assessed by following the kinetics of fungal clearance from the vagina. Rats treated with the chimeric IgG exhibited faster *C. albicans* elimination from the vagina when compared to control groups, and the CFU values were lower in vaginal fluids at all time points tested.

11 West Nile mAb

West Nile Virus (WNV) outbreaks have occurred in Africa, Asia, the Middle East and Europe. Recently, it has spread to North America where more than 29,000 cases have now been diagnosed. There is a lack of effective and specific antiviral treatment for WNV. Hu-E16 is a humanized antibody that binds to a conserved epitope on the virus envelope protein (Lai et al. 2010). Despite the promise that Hu-E16 and other mAbs have as prophylactics and therapeutics for WNV or other infectious diseases, their application may be limited by the high production costs and scalability associated with the mammalian-cell culture production system. Moreover, if biological drugs are too costly to produce for resource-poor health care systems, their therapeutic potential may never be realized.

11.1 Industrialization

To ensure high-level expression in Nicotiana, the coding sequences of Hu-E16 light chain (LC) and heavy chain (HC) were optimized in silico with *N. benthamiana*-optimized codons, and expressed using the magnICON system.

pHu-E16 was extracted and purified by a three-step purification protocol comprised of ammonium sulfate precipitation, protein A affinity, and DEAE-anion exchange chromatography. Precipitation with ammonium sulfate removed the most abundant plant host protein, the photosynthetic enzyme RuBisCo, and other plant proteins. Protein A affinity chromatography removed the remaining contaminating proteins and enriched pHu-E16 to greater than 95 % purity. The protocol produced highly purified pHu-E16 from *N. benthamiana* plants with consistent recovery among batches of different scale. In total, >5 g of Hu-E16 was purified from 16 kg of plant material for in vitro and in vivo studies. Comparable expression of E16 was achieved with Geminiviral replicon vectors in Lettuce (Lai et al. 2012).

11.2 Efficacy

Prophylaxis studies were performed in 5 week-old wild-type C57BL/6 mice (Lai et al. 2010). Mice were significantly protected when administered as little as 0.1 μg of pHu-E16. More than 80 % of mice were protected from lethal infection when 10 μg of pHu-E16 was administered.

For therapeutic studies (Lai et al. 2010), mice were passively administered a single dose (4–100 μg) of pHu-E16 at day 2 after subcutaneous inoculation of 10^2 pfu of WNV. Notably, 20 μg of pHu-E16 protected most mice from lethal infection when given two days after WNV inoculation, and a single injection of as low as 4 μg also prevented mortality. Because WNV spreads to the brain in mice by day 4 after infection, the efficacy of pHu-E16 at this later time point was investigated. A single administration of 50 μg of pHu-E16 protected up to 70 % of mice from lethal infection, and a 90 % survival rate was achieved with a single 500 μg dose.

12 Rabies mAbs

Rabies is an acute viral disease transmitted through the bite of rabid animals. There is no proven treatment once the symptoms have developed and if post exposure treatment is not administered, the disease is fatal, resulting in about 55,000 deaths per year in countries such as Asia and Africa. Treatment consists of washing the wound, administering a rabies immunoglobulin (RIG) into the wound and muscle, and vaccination. Currently, RIG from either human or horse (equine) blood plasma is used for post exposure treatment and both products carry certain risks. In the case of the equine product, administration of the serum can cause anaphylactic shock. High cost, erratic supply and the risk of blood-borne infections is associated with the human RIG. A replacement for existing RIGs is needed.

12.1 Industrialization

In a project led by CSIR (Pretoria, South Africa), two humanized anti-rabies mAbs (CSIR 1, CSIR 2) were produced using the magnICON system. Scale-up to 30 kg batches was performed at Kentucky BioProcessing.

12.2 *Efficacy*

Similar to their mammalian expressed counterparts, the mAbs were shown to be potent neutralizers and protected hamsters in a lethal challenge model and were superior in their activity when compared with human RIG.

13 Multipurpose Antibody Microbicides for Sexual and Reproductive Health

Multipurpose Prevention Technologies (MPTs) are being explored via combinations of microbicides, contraceptives and microbicides, devices and microbicides, microbicides and vaccines, and other novel approaches (CAMI, 2011). [Multipurpose products are also priorities in biodefense–see BARDA strategic plans, 2011–2016)]. Because of their high potency, specificity, and safety profile, mAbs are leading candidates for MPTs (Cone and Whaley, 1994).

Broadly neutralizing HIV antibodies 2F5, 2G12, 4E10 manufactured in mammalian cells and combined as MabGel have completed early clinical trials as a vaginal microbicide (Morris et al. 2010). The cost and capacity challenges for an antibody-based microbicide are significant–a 10 mg multipurpose microbicide if used twice a week by ten million women would require $\sim$10,000 kg of mAb per year. Coupled with conventional unit operations for very large scale mAb purification (Kelly 2007), the COG/dose could be on the order of $0.10 US dollar for *Nicotiana*–based manufacturing. MagnICON expressed HIV mAb 2G12 has been produced in the ΔXF *Nicotiana* line, and was found to contain a relatively homogeneous N-glycan species without detectable xylose or α-1,3-fucose residues (Strasser et al. 2008). Further, fully galactosylated 4E10 and 2G12 were reported to be several fold higher in neutralization potency than CHO produced mAbs (Strasser et al. 2009) whereas sialylated 2G12 exhibits similar in vitro HIV neutralization potency to other glycoforms derived from plants and CHO cells (Castilho et al. 2010).

A vaginal microbicide containing Nicotiana–derived mAbs against HIV and HSV (mapp66) is in preclinical and clinical development. Future generations of mapp66 may include additional protective mAbs against cell-associated HIV and other STI pathogens (e.g., Chlamydia, *N. gonorrhoeae*).

13.1 *Industrialization*

Antibodies for mapp66 (VRC01-N, 4E10-N, HSV8-N) have been produced (stage-appropriate GMP) in large batches (600 kg of plant) for primate studies. Yield and purities are shown in Table 2.

Table 2 Yield* and purities of mAbs for mapp66, a multipurpose (HSV/HIV) microbicide

mAb	Protein A Yield (mg/kg)	CHT** Yield (mg/kg)	Purity (%)	% Aggregate by HPLC-SEC
HSV8	79–126	23–87	97	1
VRC01	60–125	40–79	99	<1
4E10	52–104	25–47	99	<1

* mg of purified mAb/kg of biomass
** CHT ceramic hydroxyapatite

Similar yields (~ 100 mg/kg) have been achieved for 2G12 (anti-HIV) with the non-replicating CPMV-HT system (Sainsbury, 2010). The mAb displayed mostly oligo-mannosidic type glycans with Man-7 and Man-8 predominating when KDEL tagged for ER retention.

13.2 Efficacy

Currently the HIV mAbs are being evaluated for efficacy in vivo using a rigorous SHIV challenge model (repeated low dose challenge: 1 ml 10 $TCID_{50}$ of a well-characterized SHIV162p3 viral inoculum). Titrations of the mAbs are being performed sequentially starting with the highest 20 mg mAb dose delivered intravaginally in HEC gel (30 min before challenge) against 5 consecutive low dose SHIV162p3 challenges delivered at weekly intervals. Should the treated monkeys remain uninfected after these 5 challenges, the mAb dose will be lowered 4-fold and the challenges repeated 5 times, until a mAb dose is reached at which the animals are no longer protected vaginally.

14 Summary

This review summarizes the progress with large scale manufacturing of mAbs using plant viral transient systems that have advanced products to evaluation in animal models or clinical trials. Many of these studies include some data regarding purification yield. We suggest that the plant-derived mAb field has matured sufficiently that reporting mg/kg of purified mAb (or other protein) should be *de rigeur* for publications on plant virus-based manufacturing (as opposed to, or in addition to reporting expression levels in tissue).

The mAbs described herein are part of a new generation of mAb-based products that have the potential to alter the pharmacoeconomic landscape due to the benefits of transient, plant virus-based manufacturing in *Nicotiana*. These benefits include the plant virus platform's versatility, speed, scalability, low cost, and flexibility with respect to desired glycoforms. These attributes (Table 3) may have a

Table 3 Critical path dimensions for emerging antibody-based products

Target (sponsor)	Safety and/or efficacy demonstration	Maximum efficacy demonstrated	Plant virus expression system(s)	Predominant glycans	Yield post-protein A (mg/kg)*	Production scale (kg of plants)
Breast Cancer (Icon)	Mice	80% reduction in tumor volume	magnICON	Complex plant	200–300	Not reported
Candida	Mice, Rat	80% reduction in bacterial load	CAMV/TMV	Complex plant	50	Not reported
Ebola/mAbs (Mapp/KBP)	Mice Primate	100% survival	magnICON	GnGn	26–164	500
Ebola/Immune Complexes Vaccine (ASU)	Mice	80% survival	Gemini	Complex plant	2–3<	1–1.5
HSV/HIV Multipurpose Microbicide, (Mapp/KBP)	Primate	Currently being tested	magnICON	GnGn	104–126	600
NHL (Icon)	Mice, Humans	17% survival**	magnICON	Complex plant	10–200	5
Rabies (CSIR)	Hamsters	>human rabies immunoglobulin	magnICON	GnGn	62–128	30
Ricin (Mapp)	Mice	100 % survival	magnICON	GnGn	7	5
RSV (Mapp)	Cotton rat	2–3 $\log_{10}$ reduction in lung viral titer	magnICON	GnGn	180	2
West Nile (ASU)	Mice	90% survival	magnICON Gemini	GnGn	310	5

* Level of optimization of expression and purification is unknown. In the authors' facilities, optimization can result in dramatic improvements in expression and yield. ** Time to death was also significantly delayed in animals that succumbed.

significant impact on the development of multipurpose products intended for mucosal or systemic applications.

Acknowledgments Some of the work described was supported by grant numbers U19AI096398, AI62150, AI061270, AI063681, from NIAID, DAMD 17-02-2-0015 from the department of defense, and partially supported by the defense threat reduction agency grant no. 4.10007-08-RD-B. The content is solely the responsibility of the authors and does not necessarily represent the official views of the national institutes of health, U.S. Army, or department of defense.

References

BARDA Strategic plan 2011–2016

Bendandi M (2006) Clinical benefit of idiotype vaccines: too many trials for a clever demonstration. Rev Recent Clin Trials 1:67–74

Bendandi M (2008) Aiming at a curative strategy for follicular lymphoma. CA Cancer J Clin 58:305–317

Bendandi M (2009) Idiotype vaccines for lymphoma: prof-of-principles and clinical trial failures. Nat Rev Cancer 9:675–681

Bendandi M, Marillonnet S, Kandzia R et al (2010) Rapid, high-yield production in plants of individualized idiotype vaccines for non-Hodgkin's lymphoma. Ann Oncol 21(12):2420–2427

Bendandi M, Becerra C et al (2011a) Idiotypic vaccination of patients with relapsed follicular lymphoma using a novel vaccine formulation including a tobacco plant-produced tumor specific idiotype. Blood (ASH annual meeting abstracts) 118:1649

Bendandi et al (2011b) Single agent bendamustine is an effective pre-vaccine treatment for patients with relapsed follicular lymphoma undergoing idiotypic vaccination blood (ASH annual meeting abstracts), 118: 2691

CAMI (2011) Multipurpose prevention technologies for reproductive health. Report of a symposium, Washington, DC

Capodicasa C, Chiani P, Bromuro C et al (2011) Plant production of anti-glucan antibodies for immunotherapy of fungal infections in humans. Plant Biotech J 9:776–787

Castilho A, Strasser R, Stadlmann J et al (2010) In planta protein sialylation through overexpression of the respective mammalian pathway. J Biol Chem 285:15923–15930

Castilho A, Bohorova N, Grass J, Bohorov O, Zeitlin L, Whaley K, Altmann F, Steinkellner H (2011) Rapid high yield production of different glycoforms of ebola virus monoclonal antibody. PLoS One 6:e26040

Chargelegue D, Drake PMW, Obregon P et al (2005) Highly immunogenic and protective recombinant vaccine candidate expressed in transgenic plant. Infect Immun 73:5915–5922

Cone RA, Whaley KJ (1994) Monoclonal antibodies for reproductive health: preventing sexual transmission of disease and pregnancy with topically applied antibodies. Am J Reprod Immunol 32:114–131

De Muynck B, Navarre C, Boutry M (2010) Production of antibodies in plants: status after twenty years. Plant Biotech J 8:529–563

Falsey AR, Walsh EE (2000) Respiratory syncytial virus infection in adults. Clin Microbiol Rev 13:371–384

FDA (2004) Challenge and opportunity on the critical path to new medical products, food and drug administration

Giritch A, Marillonnet S, Engler C et al (2006) Rapid high-yield expression of full-size IgG antibodies in plants coinfected with noncompeting viral vectors. PNAS 103:14701–14706

Gomord V, Fitchette AC, Menu Bouaouiche L et al (2010) Plant-specific glycosylation patterns in the context of therapeutic protein production. Plant Biotech J 8:564–587

Han LL, Alexander JP, Anderson LJ (1999) Respiratory syncytial virus pneumonia among the elderly: an assessment of disease burden. J Infect Dis 179:25–30

Huang Z, Phoolcharoen W, Lai H et al (2010) High-level rapid production of full-size monoclonal antibodies in plants by a single-vector DNA replicon system. Biotechnol Bioeng 106:9–17

Hudis CA (2007) Trastuzumab—mechanism of action and use in clinical practice. N Engl J Med 357:39–51

Hudziak RM et al (1989) p185HER2 monoclonal antibody has antiproliferative effects in vitro and sensitizes human breast tumor cells to tumor necrosis factor. Mol Cell Biol 3:1165–1172

Joffe S, Ray GT, Escobar GJ et al (1999) Cost-effectiveness of respiratory syncytial virus prophylaxis among preterm infants. Pediatr 104:419–427

Kelley B (2007) Very large scale monoclonal antibody purification: the case for conventional unit operations. Biotechnol Prog 23:995–1008

Klimyuk V et al. (2012) Production of recombinant antigens and antibodies in nicotiana benthamiana using magnifection technology: GMP-compliant facilities for Small- and Large-Scale Manufacturing (THIS BOOK)

Komarova TS, Kosorukov SV et al (2011) Plant-made trastuzumab (herceptin) inhibits HER2/Neu+ cell proliferation and retards tumor growth. PLoS One 6:e17541

Lai H, Engle M, Fuchs A et al (2010) Monoclonal antibody produced in plants efficiently treats west nile virus infection in mice. PNAS 107:2419–2424

Lai H, He J, Engle M et al (2012) Robust production of virus-like particles and monoclonal antibodies with geminiviral replicon vectors in lettuce. Plant Biotech J 10:95–104

Lopez-Diaz CA, Inoges S (2009) Future of idiotypic vaccination for B-cell lymphoma. Expert Rev Vaccines 8:43–50

McCormick AA, Reinl SJ et al (2003) Individualized human scFv vaccines produced in plants: humoral anti-idiotype responses in vaccinated mice confirm relevance to the tumor Ig. J Immunol Methods 278:95–104

McCormick AA, Reddy S et al (2008) Plant-produced idiotype vaccines for the treatment of non-Hodgkin's lymphoma: safety and immunogenicity in a phase I clinical study. PNAS 105:10131–10136

Morris G et al (2010) A prospective randomized double blind placebo-controlled phase 1 pharmacokinetic and safety study of a vaginal microbicide gel containing three potent broadly neutralizing monoclonal antibodies (2F5, 2G12, 4E10) (MabGel). Microbicides (Abstract LB1)

Neal LM et al (2010) A monoclonal immunoglobulin G antibody directed against an immunodominant linear epitope on the ricin A chain confers systemic and mucosal immunity to ricin. Infect Immun 78:552–61

Niwa R et al (2004) Defucosylated chimeric anti-CC chemokine receptor 4 IgGl with enhanced antibody-dependent cellular cytotoxicity shows potent therapeutic activity to T-cell leukemia and lymphoma. Cancer Res 64:2127–2133

O'Hara JM, Whaley KJ, Pauly M (2012) Plant-based expression of a partially humanized neutralizing monoclonal IgG directed against an immunodominant epitope on the ricin toxin A subunit. Vaccine 30:1239–1243

O'Hara JM, Neal LM, McCarthy EA et al (2010) Folding domains within the ricin toxin A subunit as targets of protective antibodies. Vaccine 28:7035–7046

Olsnes S (2004) The history of ricin, abrin and related toxins. Toxicon 44:361–370

Phoolcharoen W, Bhoo SH, Lai H, Ma J, Arntzen CJ, Chen Q, Mason HS (2011a) Expression of an immunogenic ebola immune complex in nicotiana benthamiana. Plant Biotech J 9:807–816

Phoolcharoen W, Dye JM, Kilbourne J, Piensook K, Pratt WD, Arntzen CJ, Chen Q, Mason HS, Herbst-Kralovetz MM (2011b) A nonreplicating subunit vaccine protects mice against lethal Ebola virus challenge. PNAS 108:20695–20700

Piccart-Gebhart MJ, Procter M, Leyland-Jones B et al (2005) Trastuzumab after adjuvant chemotherapy in HER2-positive breast cancer. N Engl J Med 353:1659–1672

Pogue GP, Vojdani F, Palmer KE et al (2010) Production of pharmaceutical-grade recombinant aprotinin and a monoclonal antibody product using plant-based transient expression systems. Plant Biotech J 8:638–654
Roeckl-Wiedmann I, Liese JG, Grill E et al (2003) Economic evaluation of possible prevention of RSV-related hospitalizations in premature infants in Germany. Eur J Pediatr 162:237–244
Sainsbury F, Lomonossoff GP (2008) Extremely high-level and rapid protein production in plants without the use of viral replication. Plant Physiol 148:1212–1218
Sainsbury F, Sack M, Stadlmann J et al (2010) Rapid transient production in plants by replicating and non-replicating vectors yields high quality functional anti-HIV antibody. PLoS One 5:1–10
Schahs M, Strasser R, Stadlmann J et al (2007) Production of a monoclonal antibody in plants with a humanized N-glycosylation pattern. Plant Biotech J 5:657–663
Shields RL et al (2002) Lack of fucose on human IgGl N-linked oligosaccharide improves binding to human FcyRIII and antibody-dependent cellular toxicity. J Biol Chem 277:26733–26740
Shinkawa T et al (2003) The absence of fucose but not the presence of galactose or bisecting N-acetylglucosamine of human IgGl complex-type oligosaccharides shows the critical role of enhancing antibody-dependent cellular cytotoxicity. J Biol Chem 278:3466–3473
Slamon DJ, Leyland-Jones B, Shak S et al (2001) Use of chemotherapy plus a monoclonal antibody against HER2 for metastatic breast cancer that overexpresses HER2. N Engl J Med 344:783–792
Smith I, Procter M, Gelber RD et al (2007) Two year follow-up of trastuzumab after adjuvant chemotherapy in HER2-positive breast cancer: a randomized controlled trial. Lancet 369:29–36
Stirpe F (2004) Ribosome-inactivating proteins. Toxicon 15:371–383
Strasser R, Stadlmann J, Schahs M et al (2008) Generation of glyco-engineered nicotiana benthamiana for the production of monoclonal antibodies with a homogeneous human-like N-glycan structure. Plant Biotech J 6:392–402
Strasser R, Castilho A, Stadlmann J et al (2009) Improved virus neutralization by plant-produced anti-HIV antibodies with a homogeneous beta1,4-galactosylated N-glycan profile. J Biol Chem 284:20479–20485
Vezina L-P, Faye L, Lerouge P et al (2009) Transient co-expression for fast and high-yield production of antibodies with human-like N-glycans in plants. Plant Biotech J 7:442–455
Vogel AM, McKinlay MJ, Ashton T et al (2002) Cost-effectiveness of palivizumab in New Zealand. J Paediatr Child Health 38:352–357
Werner S, Marillonnet S, Hause G et al (2006) Immunoabsorbent nanoparticles based on a tobamorvirus displaying protein A. PNAS 103:17678–17717
Whaley KJ, Hiatt A, Zeitlin A (2011) Emerging antibody products and nicotiana manufacturing. Hum Vaccines 7:349–356
Wilson JA, Hevey M, Bakken R et al (2000) Epitopes involved in antibody-mediated protection from ebola virus. Science 287:1664–1666
Yermakova A, Mantis NJ (2011) Protective immunity to ricin toxin conferred by antibodies against the toxin's binding subunit (RTB). Vaccine 29:7925–7935
Zeitlin L, Pettitt J, Scully C et al (2011) Enhanced potency of a fucose-free monoclonal antibody being developed as an ebola virus immunoprotectant. PNAS 108:20690–20694
Zeitlin L et al (2012) Prophylactic and therapeutic testing of nicotiana-derived anti-RSV human monoclonal antibodies in the cotton rat model. (Submitted)

Production of Recombinant Antigens and Antibodies in *Nicotiana benthamiana* Using 'Magnifection' Technology: GMP-Compliant Facilities for Small- and Large-Scale Manufacturing

Victor Klimyuk, Gregory Pogue, Stefan Herz, John Butler and Hugh Haydon

Abstract This review describes the adaptation of the plant virus-based transient expression system, magnICON® for the at-scale manufacturing of pharmaceutical proteins. The system utilizes so-called "deconstructed" viral vectors that rely on *Agrobacterium*-mediated systemic delivery into the plant cells for recombinant protein production. The system is also suitable for production of hetero-oligomeric proteins like immunoglobulins. By taking advantage of well established R&D tools for optimizing the expression of protein of interest using this system, product concepts can reach the manufacturing stage in highly competitive time periods. At the manufacturing stage, the system offers many remarkable features including rapid production cycles, high product yield, virtually unlimited scale-up potential, and flexibility for different manufacturing schemes. The magnICON system has been successfully adaptated to very different logistical manufacturing formats: (1) speedy production of multiple small batches of individualized pharmaceuticals proteins (e.g. antigens comprising individualized vaccines to treat NonHodgkin's Lymphoma patients) and (2) large-scale production of other pharmaceutical proteins such as therapeutic antibodies. General descriptions of the prototype GMP-compliant manufacturing processes and facilities for the product formats that are in preclinical and clinical testing are provided.

V. Klimyuk (✉) · S. Herz
Icon Genetics, Weinbergweg 22, 06120 Halle (Saale), Germany
e-mail: klimyuk@icongenetics.de

G. Pogue · H. Haydon
Kentucky BioProcessing, LLC, Owensboro, KY, USA

J. Butler
Bayer Innovation GmbH, Merowinger Platz 1, 40225 Düsseldorf, Germany

G. Pogue
The IC2 Institute, University of Texas at Austin, Austin, TX, USA

Current Topics in Microbiology and Immunology (2012) 375: 127–154
DOI: 10.1007/82_2012_212

Published Online: 15 April 2012

Contents

1 Introduction

In recent years, many groups are selecting plants as the production hosts for recombinant proteins. The reasons are obvious—plants are free of animal and human pathogens, provide eukaryotic type of posttranslational modifications and offer virtually unlimited scalability for protein production. There are numerous comprehensive reviews describing different approaches for the production of recombinant proteins in plants (Knäblein and McCaman 2003; Ma et al. 2003; Fischer et al. 2004; Hellwig et al. 2004; Horn et al. 2004; Stoger, et al. 2004; Gleba et al. 2005, 2007; Klimyuk et al. 2005; Lienard et al. 2007; Decker and Reski 2008; Mett et al. 2008; Ko et al. 2009; Rybicki 2009; Vezina et al. 2009; De Muynck et al. 2010; Pogue et al. 2010; Gleba and Giritch 2011). These publications address economic, technical and safety aspects of plants as production hosts, including scalability of plant-based production, purification costs, yield, engineering of desired glycosylation pattern, and other topics.

To date, only three plant-made pharmaceutical proteins have progressed through clinical testing into the actual marketing stage of a product: anti-dental caries antibody (Planet Biotechnology); Newcastle virus vaccine (Dow Agro-Sciences) and glucocerebrosidase (Protalix Biotherapeutics). All three products have been developed on the basis of first generation technology (stable nuclear transformation): tobacco cell suspension for Newcastle virus vaccine, transgenic tobacco plants for anti-caries antibody (Weintraub et al. 2005; Wycoff 2005), and carrot cell suspension for glucocerebrosidase (Aviezer et al. 2009). These examples, despite the use of first generation technology, represent important proof of

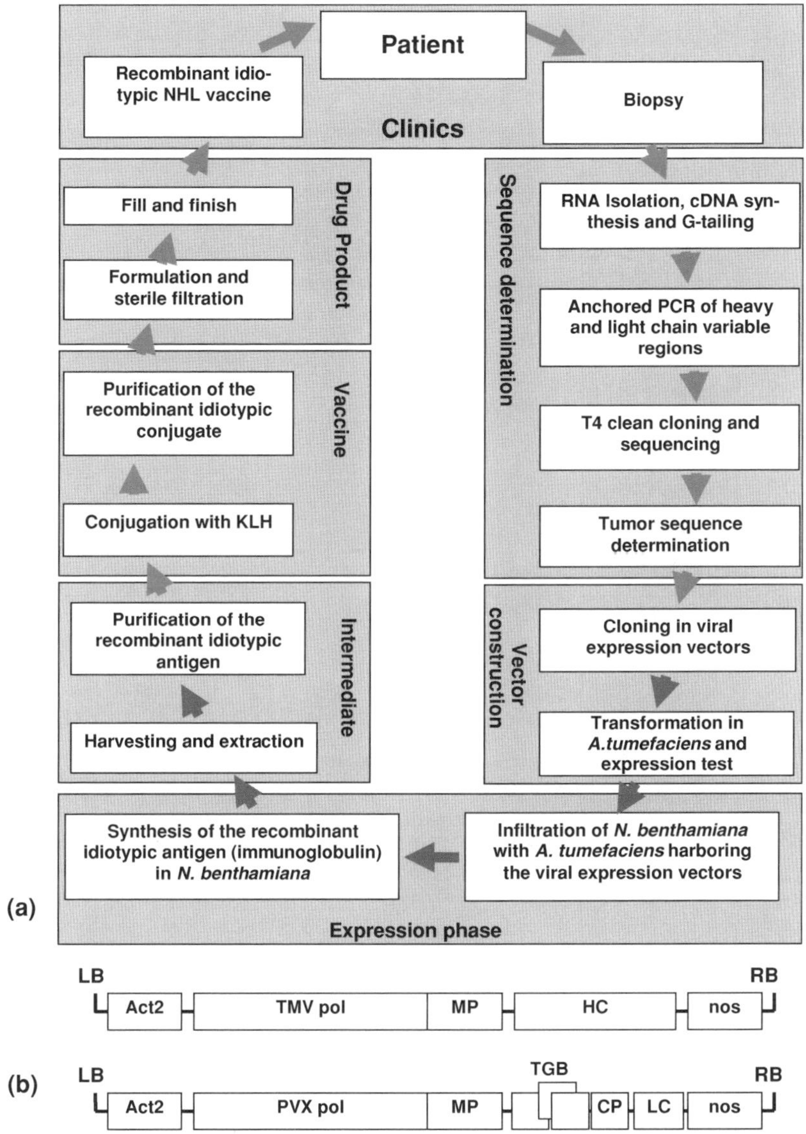

Fig. 1 **a** Schematic presentation of NHL vaccine production process. Intermediate sequence determination and vector construction are integral parts of molecular biology process. The cGMP process starts at the stage of harvesting and extraction (intermediate), includes steps of the conjugate preparation and purification (vaccine) and ends after fill and finish (drug product). **b** magnICON® vectors structure used for the expression of heavy (HC) and light (LC) chains of idiotypic immunoglobulin. Act2—transcriptional promoter of *A. thaliana ACTIN2* gene. LB, RB—left and right borders of T-DNA; TMV POL—RNA-dependent RNA polymerase of Tobacco Mosaic Virus coding region; PVX POL—RNA-dependent RNA polymerase of Potato Virus X coding region; TGB—Triple Gene Block coding region; CP—viral Coat Protein coding region; MP—viral Movement Protein coding region; nos—nopaline synthase transcription termination region

Table 1 Characteristics of different *N. benthamiana*-based transient expression systems

Vector	Protein	Expression (g/kg of FW leaves)	Time (DPI)	Agrobacterium concentration in infiltration media (OD_{600})	Expression of hetero-oligomeric proteins	Options	Reference
Plastocyanin	Murine IgG C5-1	0.2–1.5	6	0.24–0.64	Yes	+HcPro	Vezina et al. 2009
TMV	GFP (inoculated)	0.6–1.2	4	0.02	No	+P19	Lindbo 2007a
	GFP (systemic)	0.6–1.2	10–14				
TMV	GFP	3.3–5.5	5–6	0.01–0.003	No		Lindbo 2007b
TMV	GFP	4.0	6-8	0.0035	No		Marillonnet et al. 2005
TMV/PVX	23 different IgG	0.5–4.7	6	0.002	Yes		Giritch, Marillonnet et al. 2006; Bendandi et al. 2010
BeYDV	IgG 6D8	0.5	4	0.2	Yes	+P19	Huang et al. 2009
BeYDV	L1 and p24	0.55 and 0.003	3–7	0.25	No	+NSs	Regnard, Halley-Stott et al. 2010
CPMV	GFP, DsRed, HbcAg, IgG 2G12	1.6, 1.6, 1, 0.325	6	1.2	Yes	+P19	Sainsbury and Lomonossoff 2008

principle for plant-made pharmaceutical proteins production, fit into established regulatory compliance and acceptance systems, and, thus lower the risks associated with plants as manufacturing hosts for other products.

The need for speed and adaptable scale turned most groups from first generation technologies to use transient expression systems in nonfood, nonfeed plants. Among nonfood plants *Nicotiana benthamiana* has became the prodution host of choice, predominantly due to its susceptibility to different plant viral vectors (Goodin et al. 2008), and adaptation to production environments (Pogue et al. 2002, 2010). This plant is also convenient host for transient expression using nonviral vectors in combination with supressors of the transgene silencing (Johansen and Carrington 2001; Mallory et al. 2002; Voinnet et al. 2003; Vezina et al. 2009). Most new product candidates in preclinical or clinical trials are produced using *N. benthamiana* as host and transient expression systems [for review, see Gleba and Giritch (2011)]. This consistency addresses perceived regulatory and technical challenges for pharmaceuticals production in plants by excluding an inherent risk of food supply contamination, while positively addressing public concerns of using plants as production host by providing fast, highly efficient, and thus economically competitive way of recombinant protein production.

This review provides comparative analysis of the transient expression systems using *N. benthamiana* as the production host and the description of industrial prototypes for one of the most advanced among such systems based on magnICON® technology.

2 Plant-Based Transient Expression Systems and magnICON® Technology

2.1 Comparative Analysis of Different Transient Expression Systems Using N. benthamiana *as Production Host*

Transient expression (Fischer et al. 1999) is a fast and convenient alternative to stable transformation because of the speed and yield it offers for product research and development. There are two types of transient expression system, depending on the expression vectors used. The one is based on standard (nonviral) vectors comprised of the coding sequence of interest under transcriptional control of strong constitutive promoters (Vaquero et al. 1999; D'Aoust et al. 2008, 2009; Vezina et al. 2009). The second type utilizes plant viruses, predominantly the RNA viruses, as adapted expression vectors. These virus-based systems can be divided in two subgroups: vectors built on the basis of complete virus (Dawson et al. 1986; Donson et al. 1991; Turpen et al. 1993; Usha et al. 1993; Pogue et al. 2002; Lindbo 2007a, b; Sainsbury et al. 2008) and vectors built on "deconstructed" virus (Gleba et al. 2004, 2005, 2007; Marillonnet et al. 2004; Marillonnet et al. 2005; Giritch et al. 2006; Santi et al. 2006; Sainsbury and Lomonossoff 2008; Bendandi et al.

2010; Gleba and Giritch 2011). The most recent review addressing different specifics of transient expression technology are described by (Gleba and Giritch 2011).

In the following section, the versatility of each system for expressing different (size and complexity of the structure) recombinant proteins will be evaluated, as well as, the suitability of each system for industrial applications.

Comparative analysis of different *N. benthamina*-based transient expression systems is summarized in Table 1. The parameters for comparison were selected taking into account the ability of the system to express variety of recombinant proteins, especially hetero-oligomeric candidates, expression level, and other features important for scalability of the system in industrial applications.

As an example of nonviral expression system for transient expression is the system successfully used by Medicago, Inc. (Vezina et al. 2009). Strong constitutive promoter of plastocyanin gene from *Medicago sativa sp* was employed for constructing expression cassettes. Plastocyanin-based transient expression of the gene of interest, in combination with the potyviral HcPro suppressor of post-transcriptional gene silencing (PTGS), provided the accumulation of up to 1.5 g of antibody per kg of syringe-infiltrated fresh weight (FW) leaf biomass (on average 0.6–0.7 g/kg of FW). Results for the vacuum-infiltrated leaves harvested with petioles were ca 0.2–0.3 g/kg of FW. The concentration of *Agrobacterium* in the agroinfiltration medium was ca. 0.5 OD_{600}. The highest yield was achieved when the ER-retention signal, a KDEL peptide, was added to the C terminus of the immunoglobulin heavy chain.

A tobacco mosaic virus (TMV) vector adopted for agroinfiltration in *N. benthamiana* was described by J. Lindbo. Two variants of the vector were developed—one employed the full TMV virus capable of systemic movement (Lindbo 2007a), and a second—lacked the virus coat protein and was restricted with regards to systemic movement (Lindbo 2007b). Significantly higher marker gene accumulation (green fluorescent protein—GFP; up to 5.5 mg/g FW) was achieved using the vector lacking CP gene (called TRBO vector). This vector did not require the coexpression of a PTGS suppressor and worked with very high dilution rate of agrobacteria (up to 0.003 OD_{600}) in infiltration medium.

A single component vector DNA replicon system derived from bean yellow dwarf geminivirus (BeYDV) was used to to express immunoglobulin proteins (Huang et al. 2009). The BeYDV-derived replicon promotes mAb accumulation level up to 0.5 g/kg FW 4 days after infiltration. The system requires coexpression of the PTGS suppressor P19 and concentration of Agrobacteria 0.2 OD_{600} for plants infiltration. The BeYDV-derived vector system was also used by the Edward Rybicki group (Regnard et al. 2010) to express the L1 protein of human papilloma virus HPV-16 and the p24 protein of HIV-1. Similar to the above-mentioned case, expression of L1 and/or p24 in the geminiviral system required the use of a coexpressed PTGS suppressor (Takeda et al. 2002).

The system based on disabled RNA2 of cowpea mosaic virus (CPMV) was developed in the laboratory of George Lomonossoff. The system overcomes the historic limitation of the insert size in CPMV vectors and allows expression hetero-oligomeric proteins from a single vector (Sainsbury and Lomonossoff 2008; Sainsbury et al. 2009). However, the system also requires for the infiltration the use of relatively high concentration of agrobacterial suspension and coexpression of PTGS suppressors.

The most advanced system launched by agrobacterial infiltration is the transient system developed by Icon Genetics, Halle, Germany trademarked as magnICON® technology, and vectors. The magnICON® technology is currently used by many business entities and research laboratories worldwide. The technology and its applications have been described in numerous publications (Gleba et al. 2004, 2005, 2007; Marillonnet et al. 2004, 2005; Gils et al. 2005; Giritch et al. 2006; Huang et al. 2006; Santi et al. 2006, 2008; Vancanneyt et al. 2009; Webster et al. 2009; Bendandi et al. 2010; Gleba and Giritch 2011). The magnICON® system has produced numerous heterologous proteins, including cytokines, interferon, bacterial and viral antigens, growth hormone, single chain antibodies, and monoclonal antibodies (mAbs) at high levels in plants, 1–5 g/kg fresh weight plant material. These levels support economically viable products ranging from pharmaceutical and diagnostic analytes, to tissue culture excipients and biochemical reagents. For example, the accumulation level of immunoglobulins can be up to 4.7 g/kg of infiltrated leaf biomass (Bendandi et al. 2010). The magnICON system is simple to use, does not require coexpression of PTGS suppressors and requires only a fraction of the *Agrobacterium* vector for infiltration as other systems. A 1,000x dilution of an overnight *Agrobacterium* culture or ca. 0.002 OD_{600} is required for efficient infiltration, compared with at least 0.2 OD_{600} or 100x higher for the systems that require the coinfiltration and expression of PTGS suppressors genes. This feature is extremely important for the large-scale production when the tons of plant biomass have to be treated, as 1 L of overnight agrobacterial culture yields 1000 L (or 1 m^3) of infiltration medium, thus reducing the complexity and costs associated with the plant infiltration step in production.

Taking into account the virtual unlimited scalability of plant-based production (each plant is small "bioreactor" or "fermentor," so different production scales differ only in the plurality of plants-"bioreactors" used), the scale up procedures are relatively straightforward compared with the complexities associated with standard fermentation scale up leading to changes in product quality manufactured at different scales (Mack 2008).

In the following sections we describe the industrialization of the magnICON production system for different products requiring vastly different scales of production. The magnICON® technology does not share the major shortcomings of other plant-based technologies. It provides best combination of following features: rapid and low cost R&D; high expression levels approaching theoretical limits of the biological systems employed, ability to express complex hetero-oligomeric proteins, low biosafety concerns, and reduced production costs. The process has been brought to a current good manufacturing processs (cGMP) compliance level at Bayer using its new pilot manufacturing facility in Halle, Germany. Further, the system has been brought into cGMP compliance in second facility with a larger capacity, of 0.8–1.2 ton plant biomass/day, in collaboration with Kentucky Bioprocessing LLC, (Owensboro, KY; Pogue et al. 2010). Modifications to the KBP facility and a new facility being constructed by G-Con LLC at College Station, TX offer even greater scale using the magnICON transient expression system and the *N. benthamiana* production host. These capacities will be >2 ton/day.

The magnICON® is being used for development of a variety of products, primarily pharmaceutical proteins, by Bayer and its licensees. Bayer's lead product candidate, individualized vaccines for the active immunotherapeutic treatment of NonHodgkin's lymphoma (NHL), is produced in a cGMP-compliant manufacturing facility in Halle entered clinical trials in January 2010 (Fig. 1a).

The descriptions of two pilot manufacturing facilities with emphasis on cGMP-compliant manufacturing processes for production of individualized vaccines for NonHodgkin's lymphoma (Icon Genetics GmbH, Halle, Germany) and anti-CCR5 mAb (Kentucky Bioprocessing LLC, Owensboro, KY) are provided in the following sections.

2.2 Production Requirements for a Virus-Based Expression System

For successful pharmaceutical protein production and cGMP compliance, a plant virus-based expression vector must maintain the genetic stability of the nonviral sequences encoding the intended product under variable environmental conditions and host defense responses. Genetic drift in nonviral sequences represents a significant problem for any production vector. The high mutation rate associated with RNA viruses can raise concerns for this type of drift (Van Vloten-Doting et al. 1985). However, investigators have evaluated the properties of various tobacco mosaic virus (TMV)-based virus vectors and observed that a lack of drift in nonviral sequences during successive passages through plants (Kearney et al. 1993, 1999). In contrast to point mutations in the nonviral sequence, significant selective advantage is observed by deletion of all or large segments of the nonviral sequences. This type of rearrangement is the most common observed in RNA virus vectors on multiple passages through host plants and has been characterized at the molecular level in TMV vectors (Rabindran and Dawson 2001). Genetic stability of this type is practically measured by the use of reverse transcriptase polymerase chain reaction (RT-PCR) with primers flanking the nonviral sequence inserted into the vector. Maintenance of the nonviral insert is indicated by the size of the RT-PCR DNA product; any spurious deletion variants will be noted by smaller DNA fragments. Another practical means of measuring vector stability is the analysis of the final protein product, where mutations in the genetic sequence resulting in amino acid substitutions would be readily detectable in the protein population by sensitive liquid chromatography mass spectrometry (LC–MS) methodologies.

For magnICON® vectors, the vectors are introduced simultaneously into most plant tissues and cells making the replication and movement cycle very short compared with early generation plant virus expression vectors that need to replicate on inoculated leaves, enter the phloem and be deposited on distal tissues for cell to cell movement and complete infection. Thus, due to the short replication cycle of magnICON® vectors, ensuring the virus sequence in the master and working cell banks (MCB and WCB, respectively) is correct and consistent is the principal check point to control for integrity of the nonviral sequence encoding the protein of interest. The WCB is employed in an Agroinfiltration process that initially introduces the virus vectors, carried by the introduced *Agrobacterium*, to many cells throughout the transfected plant. The vector then is "activated" by transcription from the T-DNA region in the inserted plasmid and following modification in the nucleus, is transported to the cytoplasm for translation and the replication cycle.

Most magnICON® vectors encode requisite proteins for cell to cell movement, including the movement (30 K) protein from *tobamovirus*-based vectors and the triple block products and coat protein for *potexvirus*-based vectors (showns as MP, TGB and CP, respectively, Fig. 1b). These proteins allow movement of the virus vector genome locally within an inoculated leaf resulting in the majority of cells being infected and becoming production sites for the desired protein product in as few as five to seven days. Aerial parts of the plant are harvested generally by 6–8 days post inoculation (dpi) and extracted for the desired product. The limited movement, the en masse introduction of homogeneous T-DNA constructs to launch the RNA replicons ensures few genetic bottlenecks and low selective pressure. Indeed, deletion or other modifications of the nonviral sequence has not been noted in magnICON® production environments (Kentucky BioProcessing, LLC, unpublished data). This represents a significant advantage of magnICON® over other independent virus vectors that must establish localized infections and move systemically throughout the plant during a production run.

3 Use of magnICON® for Production of Multiple Small Batches: Individualized Idiotype Vaccines for Non-Hodgkins Lymphoma

3.1 Individualized Idiotypic Vaccines and the Production Process Requirements

Individualized medicine can provide new and often superior efficiency compared with classical medicaments. However, personalized strategis present an extraordinary challenge to the cGMP production process. In its strictest definition, individualized medicine require a unique drug product to be developed and produced for each patient. An example for such an individualized therapy is the experimental treatment of indolent NHL patients with idiotype vaccine. Standard therapy is applied to the majority of patients involving chemotherapy follow or in conjunction with immunotherapy using anti-CD20 monoclonal antibodies. Despite these treatment options, occurence of relapse is still high (Bendandi 2008; Longo 2009). The individualized idiotypic vaccines for NHL are composed of unique cell surface immunoglobulin specific for the malignant B-cell conjugated to a strong immunogen such as keyhole limpet hemocyanin (KLH). A more detailed description of the vaccine and its development is described by Bendandi and colleagues (Bendandi et al. 2010). Unlike classical drug products, production of this individualized drug can only be initiated when the patient already developed a malignancy because genetic information from the tumor cells is required for cloning of the disease-specific immunoglobulin and production of a therapeutic vaccine. Therefore, production of such a vaccine requires several specific preconditions. It implies production of multiple small batches with similar but not identical products, each product consists of patient-specific immunoglobulin conjugated to KLH. The vaccine can then be used solely for this patient. Hence, the production system display a manageable cost of goods but be sufficiently robust to support the manufacture of a multitude of products. There is a necessity for speed in the production process for complex eukaryotic proteins to supply the patient with its drug in the correct therapeutic window. Finally, the process also must exclude cross-contamination between patient-specific batches, and, obviously, be cGMP compliant.

Transient plant expression systems fulfill these requirements: No fermenter for protein production is needed, using instead a batch of transfected plants grown in single-use trays as production hosts. Up- and down-scaling of the expression quantities can be easily achieved by changing the number of trays, thus plants, used. Spatial and temporal isolation of incubation and purification of the production batches prevent cross-contamination. Only 7 days after transfection (infiltration of the *Agrobacterium* strain containing the magnICON vector encoding the idiotypic vaccine antigen) of plant host are needed to achieve suitable level of individual immunoglobulin in infiltrated plant tissues. During expression phase the plants are grown under standard growth conditions supplemented only with light, water, and fertilizer. No animal-derived compounds are needed for the production process. Additionally, the plant host does not support the growth of human or animal-specific pathogens thus reducing the concerns of impurities in the final product. At first glance, cGMP compliance of a transient plant-based production system could be thought to pose a challenge because most cGMP-regulations for protein products are designed for drug substances derived from cultivated, stably transformed cells. Hence, the traditional terms like, for example, MCB and WCB, do not truly fit the production work flow and have to be carefully defined to avoid misunderstandings.

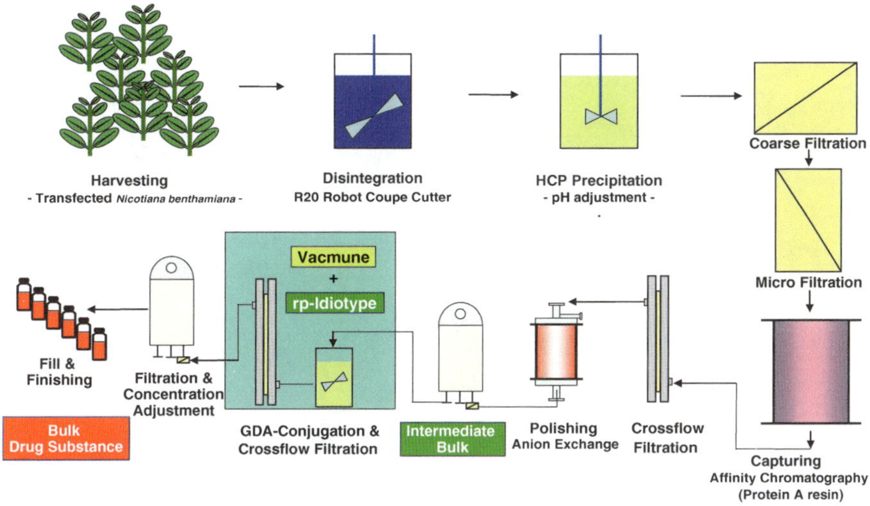

Fig. 2 Downstream-process of NHL vaccine producion

3.2 Industrial Agrobacterium Strain and Nicotiana benthamiana *Production Host*

In transient plant expression there is no stable genetically transformed plant line available which would constitute the master cell bank. Instead, master and working stocks exist on every level of the production process. As for each patient the idiotypic sequence has to be cloned into an Agrobacteria-delivered viral vector, there are controlled master and working stocks for cloning- and expression-vectors, as well as for the *Agrobacterium tumefaciens* strain and the *Nicotiana benthamiana* production host.

For industrial-scale transfection of plants using Agrobacterium, it is useful to have a well defined and well characterized *Agrobacterium* strain. Therefore, we have engineered an industrial strain, starting from the well known and entirely sequenced wild type Agrobacterium strain C58 (Goodner et al. 2001; Wood et al. 2001). Several elements were engineered in this strain. First it should be "disarmed," meaning that the Agrobacterium genes that are normally transferred to plant cells to induce cell proliferation (leading to formation of the crown gall) were deleted from transfer or T-DNA present on the wildtype Ti plasmid. Therefore, the entire T-DNA of C58, including the left and right borders, was deleted and replaced with a beta-galactosidase gene (*lacZ*) from *Escherichiacoli*. The *lacZ* gene confers a blue color when the strain grows on a plate containing the X-Gal chemical substrate. It is a useful marker to confirm that the Ti plasmid is present. Since large amount of Agrobacterium will be used, the industrial strain should be unable to grow outside of the production facility or at least compromised with regards its

Fig. 3 Clinical manufacturing facility. **a** Authomatic seeding machine. **b** Plant trays 1 week after seeding. **c** Plant tray ready for infiltration fixed in infiltration cage. **d** Custom-made stainless steel vacuum-infiltration apparatus with upside down inserted plant tray. **e** Homogenized plant material after harvesting. **f** Semi-automatic filling station (Acerta system, Millipore Corporation, USA)

"fitness." For this purpose, two auxotrophies were engineered in the strain such that they require cysteine and thiamine for growth. For cysteine, two homologous but unlinked chromosomal genes were deleted, and for thiamine, two genes at the third chromosomal location were deleted. The doubly auxotrophic strain grows as robustly as the wild type strain on rich medium but does not grow in minimal medium, lacking exogenous amino acids. All engineered regions were sequenced to make sure that the expected changes were made. In addition, Southern blot analysis using the vector

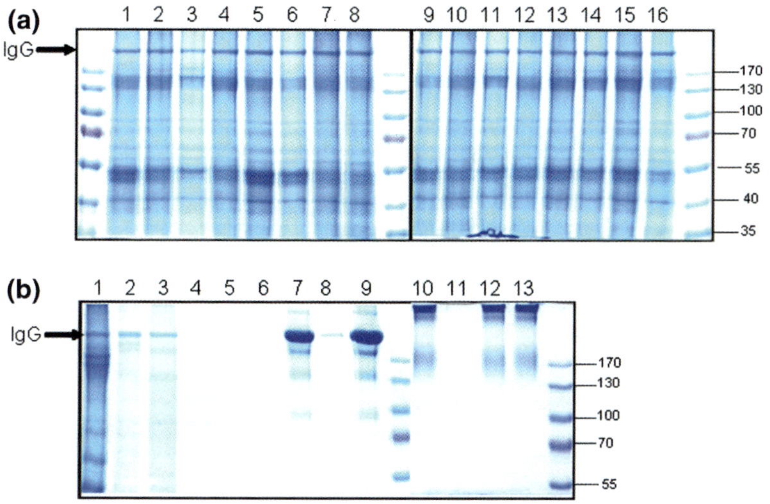

Fig. 4 **a** The SDS-PAGE analysis of the recombinant idiotypic immunoglobulins (IgG) expression in the infiltrated plant leaf tissue. The random plant samples were taken from 16 different trays (1–16). **b** The SDS-PAGE analysis of the idiotypic immunoglobulin purification and conjugate production. *1*—crude plant extract; *2*—plant extract, pH 5.0; *3*—plant extract, pH 8.5; *4*—flow-through protein A column; *5*—wash II; *6*—wash III; *7*—eluate from protein A column; *8*—UF/DF permeate (immunoglobulin); *9*—purified immunoglobulin (intermediate bulk); *10*—conjugate with KLH (immunoglobulin: KLH ratio is 2:1); *11*—UF/DF permeate (conjugate); *12*—retentate (conjugate); *13*—final bulk (conjugate)

backbone as a probe was conducted to verify that no fragments of the vector used for strain engineering had integrated at an unknown location in the genome.

It is also useful to have an antibiotic selection marker for maintaining the strain and growing it for preparation of competent cells. Rifampicin resistance was chosen as it can be easily obtained by selection for spontaneous mutation. The selected strain contains a point mutation in cluster I region of *rpoB* (ß-subunit of RNA-polymerase) similar to *E. coli* rif^R-mutants (Severinov et al. 1993). The complete genome of the resulting industrial Agrobacterium strain, designated ICF320, was sequenced, validating all introduced modifications and excluding unwanted mutations.

3.3 NHL Vaccine Production Process

From the regulatory point of view, it is also important to define an exact starting point for the cGMP-process, after which all applicable GMP-regulations are strictly applied. The production process for NHL vaccines is divided in a molecular biological process and a protein expression and purification process. The molecular biological process starts with RNA-isolation from patient's tumor biopsy material and ends with the completed expression vector transformed into the *A. tumefaciens* host. The protein expression and purification process starts with the Agrobacteria infiltration of the plant production host *N. benthamiana* and ends with filling the purified, sterile vaccine in glass vials. The vials are subsequently sealed, labeled and then shipped to the clinic. The schematic presentation of the whole process starting from biopsy collection and ending with patient treatment with idiotypic vaccine is shown in Fig. 1a. In agreement with regulatory authorities and based on the EU GMP Guide Part II "Basic Requirements for Active Substances used as Starting Materials" the starting point for the cGMP-process was set to the protein extraction step analogous to conventional active substances extracted from plant sources (see Table 1 in EU GMP Guide Part II). However, GMP-like documentation and quality control was applied to the molecular biology process, but no equipment qualification and process validation was implemented. The molecular biology process yields two specific expression vectors (one containing the gene encoding the heavy and one containing the gene encoding the light chain of the idiotypic immunoglobulin, see Fig. 1b), which are verified by sequencing to confirm desired idiotypic sequence. Agrobacterium cultivation, plant infiltration and incubation are carried out in a class D (ISO 8) clean room environment under GMP-like conditions. The actual *Agrobacterium* strain is resequenced and tested for contamination after its use in the production process to exclude unwanted mutations and potential sample transpositions. Full GMP is applied from harvesting and extracting the plant material through final fill and finish steps. Accordingly, all equipment for these stages is fully qualified and critical process like sterile filtration and filling are validated. More detailed presentation of GMP process is shown in Fig. 2.

Plant biomass, which will be used for Agrobacterium infiltration and idiotypic immunoglobulin production is grown in greenhouse compartments. Plants are grown in plastic trays on soil or mineral wool. Seeding is carried out with an automatic machine (A-35 Seed-Air-Matic, KW Automation, Australia, Fig. 3a) allowing seeding of a plant tray (Fig. 3b) in a few seconds. The plants are watered and fertilized by an automatic flooding system. Four to five weeks old plants are transferred into the clean room unit and vacuum infiltrated with a diluted Agrobacterium suspension (OD_{600}1 diluted 500 times) in a stainless steel vacuum tank (Fig. 3d). For that purpose, the plant tray is fixed in a cage (Fig. 3c) and submerged upside down in the Agrobacterium suspension. Vacuum of 80–100 m bar is applied. Rapid release of the vacuum efficiently disperses the Agrobacteria cells into the leaves resulting in a uniform and simultaneous transfection. The infiltrated plants are then transferred into an incubation chamber within the clean room facility. The plants are incubated under standard greenhouse conditions for 7 days. During this time the expression of the recombinant idiotypic immunoglobulin antigen takes place. The SDS-PAGE analysis of random plant samples from different trays show uniform and high expression of the immunoglobulin (Fig. 4a).

The process has been optimized for robustness and simplified handling, not for highest yield since each recombinant idiotypic immunoglobulin is unique and therefore exhibits different accumulation and recovery properties. A robust and reliable purification process (Fig. 2) was developed based on protein precipitation, affinity chromatography, and diafiltration (J. Lenz, personal communication).

Five kilograms of plant material (stems and leaves) per batch are harvested and disintegrated in 12 L of extraction buffer. Disintegration takes place in a vertical cutter (R20 V.V, Robot Coupe, France) commonly used in the food industry. After 5 min a viscous, green homogenate is obtained (Fig. 3e). A large fraction of host proteins including Rubisco is then removed by protein precipitation at acidic pH. The pH is subsequently adjusted to slightly alkaline conditions to facilitate antibody binding to protein-A-affinity resin. The crude extract requires clarification to remove materials that will precipitate, fibers and other particulate matter before loading on a protein-A-affinity column. This is achieved by pumping the crude extract through a serial combination of a coarse filter, micro filter and protein-A-affinity column. Single-use filter-materials and protein A affinity resin is used to prevent cross-contamination of different batches. Filter housings and column body is made of stainless steel and glass to facilitate sanitization with sodium hydroxide. Cleaning validation showed no residual protein from previous batches or cleaning agents.

The affinity column with bound antibody is then transferred into the class B (ISO 5) cleanroom area to continue the polishing steps of the antibody. The recombinant idiotypic antibody is eluted from the column and then ultrafiltrated/diafiltrated first with acidic PBS buffer and then physiological PBS buffer (pH 7.3). The first diafiltration step should remove the traces of protein A leaking from the affinity column, whereby the intermediate is adjusted to the final buffer conditions during the second diafiltration step. Subsequently, the retentate is passed through an anion exchange adsorber which is operated in negative chromatography mode. Impurities such as residual DNA are bound to the adsorber membrane, whereas the recombinant antibody flows through unbound. The antibody is then filtered through a 0.2 μm-filter into a storage bag yielding the intermediate bulk.

The average purity of the intermediate bulk is 97.4 ± 1.4%, based on the analysis of six different batches by capillary gel electrophoresis. No residual DNA or any traces of nicotine were detectable (nicotine is a typical secondary metabolite of the host plant). The endotoxin level is also below the detection limit. The protein-A level is reduced to an average level of 161 ppm.

The purified recombinant idiotypic immunoglobulin (intermediate bulk) is conjugated to a carrier protein in 1:1 (w/w) ratio to enhance the immune response. KLH is one of the strongest immunologically enhancing carrier proteins used for this purpose. A defined and stable formulation of commercially purchased, GMP-manufactured KLH subunits (VACMUNE®; biosyn Arzneimittel GmbH, Fellbach, Germany) is used as the carrier protein. The conjugate is further purified by diafiltration to yield the drug substance, recombinant idiotypic conjugate. The conjugate is again filtered through a 0.2 μm-filter into a storage bag yielding the final bulk. Size exclusion chromatography (SEC) with UV detection yields an average purity of the final bulk >99%. However, SDS-PAGE detects faint smear below the conjugate bands which virtually do not migrate into the gel due to their large MW (Fig. 4b, lines 10, 12 and 13). The final conjugate representing the idiotype vaccine consists of a mixture of recombinant tumor immunoglobulins cross-linked with glutaraldehyde to KLH subunits. The average molecular weight as determined by SEC with static light scattering detection is above 1 MDa.

The idiotype vaccine is diluted in the storage bag with PBS-buffer pH 7.3 to the desired concentration of 1.0 mg/ml. The solution is then filtered through a 0.2 μm sterilizing grade filter into a preassembled, sterile, and disposable filling module. The integrity of the filter is tested by an automatic filter-integrity test-device. Filling is carried out with a semi-automatic filling station (Acerta system, Millipore Corporation, USA; Fig. 4f) in a class A (ISO 5) laminar flow bench. All product contacting surfaces consist of single-use, GMP-compliant materials. The sterile solution is filled into glass vials under aseptic conditions. The vials are closed under aseptic conditions with stoppers and seals. After passing the release tests, the vials are shipped to the clinic for experimental treatment. In summary, the whole production process, starting from plants agro infiltration, infiltrated plants incubation period and the vaccine manufacturing is performed within 2 weeks.

4 magnICON® Scale Up and Production of Anti-CCR5 mAb

Rapid and small-scale production fits the needs of personalized products. However, most pharmaceutically active proteins are required from production lots at the multigram to multikg scale. Adaptation of the magnICON® transient expression system for large scale, multigram, bio manufacturing, environment is critical. This adaptation has been successfully accomplished using the production of a neutralizing mAb binding the CCR5 co-receptor as a case study.

4.1 Host Species and Growth Conditions

magnICON® vectors must show faithful replication of virus genomes under the less than complete control of glasshouse-grown plants. Control of temperature, humidity and airflow remains important for reproducible yields and vector performance. Extreme environmental conditions can lead to enhanced defence responses in the host plant, changing the selective landscape of a virus vector and influence the genetic stability of the vector population.

As noted in the introduction, *Nicotiana benthamiana* is used as the plant host for magnICON®-based protein production (Gleba et al. 2005, 2007). *N. benthamiana* is an indigenous species in Australia and a widely used model plant especially in plant virology (Goodin et al. 2008). This species is ideally suited for the magnICON® expression under glasshouse and controlled indoor conditions due to its susceptibility to both bacterial and viral infections. This susceptibility to *Agrobacterium* infection mediates initial entry and movement of the expression cassettes. Likewise, *N. benthamiana* is known to be almost universally susceptible

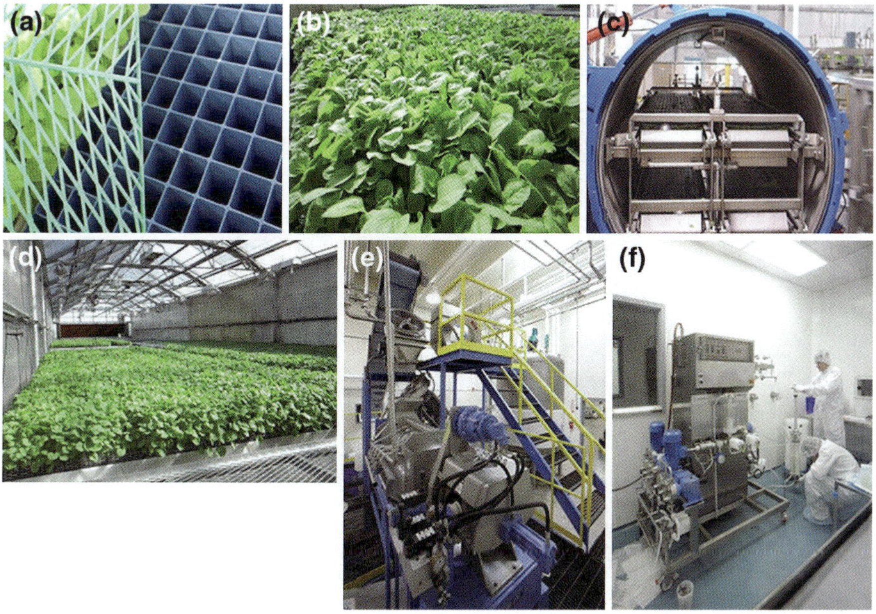

Fig. 5 "At scale" Agroinfiltration system. **a** Growth trays with specially designed lids are to seed plants containing to permit arial growth, yet provide a barrier for soil and root components. **b** Plants are grown to optimal size and density in trays. **c** On to each of four conveyors ten growth trays are loaded to enter the a vacuum-rated chamber equipped with both for and aft doors. Once loaded with trays, conveyors rotate 180° and enter the chamber. Plants are submerged in *Agrobacterium* containing solution in chamber and vacuum is applied and released. Plants are removed from chamber and rotated to upright position using conveyors. **d** Plants are transferred to greenhouses for growth and product accumulation. **e** Plants are generally cut at tray lid height and harvested materials are processed through a two-stage disintegration device and soluble plant extracts are removed from fiber using a tissue press. **f** Clarified plant extracts are subjected to various purification steps involving column chromatography, ultrafiltration and diafiltration procedures to derive final purified product

to plant viruses, partially based on a defective form of RNA dependent RNA polymerase found in its genome (Yang et al. 2004). This viral susceptibility allows external viral replicases delivered as part of the magnICON® expression system, to successfully replicate the delivered genes. The combination of ease of infection with bacterial and viral components leads to highly homogeneous, genetically stable virus replication cycles using the magnICON® vector system. Therefore, *N. benthamiana* is the host species of choice for most magnICON® glasshouse applications.

A master seed bank was produced through self-pollination of the parental plants. From this master seed bank, a working seed bank was generated starting from 80 independent mother plants. This particular seed production lot has generated an overall seed yield of 543 g, which is comprised of approximately 10^7

(i.e. 10 million) seeds. To allow biochemical and genetic analyses of all individual mother plants, leaf samples were taken and stored at −80°C. As a routine procedure all collected samples were analyzed for the potential presence of transgenes. The following commonly used transgenes were tested: 35S (CaMV), NPTII (neomycin phosphotransferase), BAR, NOS, and 2mEPSPS. In comparison to appropriate positive controls, the target genes were not detected in any of the individual mother plants confirming their nontransgenic nature. Comparative visual phenotypic assays of the plants produced from master and working seed banks were carried out under identical growing conditions. No phenotypic differences between different seed lots were detected. A new genotypic assay—amplified fragment length polymorphism (AFLP) analysis—is being developed to determine the degree of genetic variation (if any) within the different seed lots.

4.2 Large-Scale Application of magnICON® Expression Vectors

To produce full immunoglobulin proteins for therapeutic testing, larger scale preparation of production materials is essential for cGMP compliance. Therefore, the process described above must be significantly modified to support larger scale production. MCB containing magnICON® DNA vector are amplified to produce several hundred WCB vials for production purposes. Bacterial growth facilities are required varying in size from shake flasks to small multiliter bioreactors. These facilities should be kept physically separate from plant growth and other production equipment and maintained with appropriate cleaning and sterilization procedures to ensure absence of cross contamination. For a typical production run, a WCB vial is thawed and grown to ready cells for plant infection. Bacterial stocks housing magnICON® vectors can be diluted routinely 1,000-fold to obtain adequate compositions for infiltration. The infiltration process can be staged at a variety of scales, from bench top to pilot to full-scale production. Moving the infiltration process from a manual scheme to an automated system with high degree of reproducibility, the Agroinfiltration process to accommodate the infiltration of kilograms of plants per hour, allowing 25–75 g of antibody to be produced per greenhouse lot using the magnICON® vectors (automated Agroinfiltration system for large-scale production at Kentucky Bioprocessing facility is shown in Fig. 5).

For host propagatgion, the process begins with qualification of *N. benthamiana* seeds based on seed size, germination frequency and defined storage conditions and expiry dating. *N. benthamiana* seeds are placed into a tray system that accommodates 128 plants/tray in a prewetted tobacco soil mix using a precision needle seeding device (Fig. 5a). A lid is placed over the seeded soil trays such that the seedlings grow through a hole in the lid, separating the aerial portion of the plant from the soil and roots materials beneath the lid. Plants germinated in the nursery (average temperatures ranging from 20 to 35°C depending on the season) for seven days using an automatic misting system with supplemental lighting and heating/cooling. After germination, plants were moved to a production greenhouse, (average temperatures ranging from 20 to 32°C depending on the season), where they are maintained prior to Agrobacterium strain infiltration (Agroinfiltration) for ~30 days post sowing (Fig. 5b). The trays are then manually loaded onto a conveyor belt system, inverted 180° and moved into a vacuum rated autoclave (Fig. 5c). The plants are lowered into liquid reservoirs containing the *Agrobacterium* solution from the amplified WCB of a particular *Agrobacterium* strain. Cells were grown to confluence the night before and thus are fresh and metabolically active. The autoclave doors are then shut and sealed and vacuum is applied and then released to allow entrance of the *Agrobacterium* solution into the interstitial spaces of the submerged plant tissues. Infiltration of plant tissues leaves a "wetted" appearance on the leaves. On completion of vacuum cycle, trays are rotated to an upright orientation and conveyed out of the autoclave for transport to the controlled growth environment (Fig. 5d). On exit, the next set of trays is introduced and the cycle is repeated. The system has been designed to infiltrate ~750 kg of *N. benthamiana* plants in an 8 h production cycle. Plants are incubated in greenhouse environments for 7–14 days depending on product-specific optimization of plant biomass and yield. Following the defined growth period, plants were harvested by applying manual or automated blade systems cutting at the base of each infiltrated plant.

4.3 Recombinant Protein Extraction and Release Testing

Delivery and performance of the expression vector is only the start of the protein production process. The product must be efficiently extracted and purified for final use. A general strategy is described in the following. Treated plant material generally undergoes a two-step disintegration process. During the first disintegration step, the plant tissue processing equipment is configured with knives and extraction fluid containing buffering, anti-oxidant and divalent cation chelating agents at a rate equal to 1.0–2.0 L buffer per 1.0 kg of plant material. The second disintegration step continues with the plant tissue processing equipment configured with hammers to further reduce the particle size of the plant material. After disintegration, the homogenized plant material is fed into a horizontal screw press for initial separation of plant fiber from the liquid extract (Fig. 5e). As observed with the individual vaccine products, separation of plant photosynthetic membranes and other relatively insoluble, yet abundant impurities, is the next important phase in purification. Several methods are available for removal, including use of low concentrations of polyethylene glycol, filtration units, and differential centrifugation techniques. However, the most flexible approach is to modify the pH of the soluble plant extract and a high purity filter aid can be also added. Slightly elevated temperatures can be applied to acidified plant extracts resulting in precipitation of photosynthetic membranes and associated proteins. The resulting protein extract can be subjected to a variety of downstream purification methodologies such as ultrafiltration, column chromatography, filtration, and diafiltration specific for the target product.

Release testing is necessary for all recombinantly produced therapeutic candidates regardless of the production system (Stein and Webber 2001). Principal tests for release involve the measurement of the product's purity, identity, and potency. Various electrophoretic methods are used to measure the purity of the final product and the nature of protein impurities, including sodium dodecyl sulfate polyacrylamide gel electrophoresis (SDS-PAGE), iso-electric focusing gels, or 2-dimensional PAGE. Further analysis of proteinacious inhibitors can be carried out by liquid chromatography mass spectrometry (LC–MS) and matrix assisted laser desorption-ionization time of flight (MALDI-TOF) mass spectrometry (MS). Antibodies raised against the host proteins are often used to identify host plant protein impurities. Lack of other impurities, including small molecules such as nicotine from *Nicotiana* plants, or contaminants (materials introduced during biomanufacturing) are best measured by gas chromatography MS (GC–MS) and high pressure liquid chromatography (HPLC) methods. The presence of endotoxins is generally measured by the Limulus amebocyte lysate (LAL) assay per the US Pharmacopedia. Sterility assays are used to ensure the lack of microbial growth. Plants benefit from the lack of need to detect contaminating viruses since plant viruses are not known to infect mammalian cells. The concentration of contaminating DNA is measured by various testing radioactivity- and amplification-based methodologies.

The identity of the active protein can be most readily established using peptide mapping of trypsinized protein, sizing using MALDI-TOF MS methods and identity through LC–MS/MS. Immunoblotting and N-terminal amino acid sequence analysis can also provide identity testing. Immunoblotting using specific antibodies for the protein product is especially powerful for identification of the nature of breakdown products derived from the primary or full-length recombinant product. Potency is determined using assays measuring properties directly related to product mode of action including cell-based assays to provide a specific activity measurement associated with each batch of protein product (Stein and Webber 2001). Although plant viruses are nonpathogenic to humans, virus particles do represent one of the major potential impurities from plant extracts. Absence of *tobamovirus* or *potexvirus* virions in the final product is best determined using RT-PCR analysis. Primers targeting conserved regions from each virus vector can be used to amplify portions of potentially contaminating RNAs. Hypersensitivity assays can be used to detect virus vector impurities. For *tobamovirus*-based vectors produce hypersensitive responses, or local lesions, on *Nicotiana tabacum* cv. Xanthi NN plants that are quite sensitive, with detection levels at 1 ng virus/ml. End-point-dilution analysis of TMV U1 (wild-type TMV) and virus vector containing extracts can be performed on local lesion host Xanthi as well as systemic hosts *N. benthamiana*. Therefore the presence of virus vector impurities can be effectively determined using either molecular, RT-PCR, assays or functional, hypersensitivity, assays.

4.4 magnICON® Production and Characterization of Anti-CCR5 mAb

Monoclonal antibodies (mAbs) have shown broad therapeutic applications for infectious disease, oncology, inflammation, allergy, and cardiovascular areas (Hoentjen and van Bodegraven 2009; Weiner et al. 2009). In the treatment of virus infections, mAbs are being explored to block the entry of viruses into cells to prevent infection (Trkola et al. 2001; Murga et al. 2006; Shearer et al. 2006; Jacobson et al. 2008). Human immunodeficiency virus type 1 (HIV-1) represents a significant challenge to human health throughout the world. The chemokine (C-C motif) receptor-5 (CCR5) acts as a co-receptor for HIV-1 entry into cells (Moore et al. 1997). The use of an mAb to block viral access to CCR5 may serve as a strategy to prevent HIV-1 infection (Gaertner et al. 2008). Further, CCR5 appears to be nonessential for human health since individuals with CCR5-Δ32 alleles (essentially a CCR5 knock-out) are healthy (Dean et al. 1996).

magnICON® is used to express the humanized heavy and light chains of a mAb that specifically binds the ligand-binding domain of the human CCR5 receptor. magnICON® vectors are constructed from two noncompetitive virus genomes, the first based on turnip vein clearing tobamovirus (TVCV) and the second based on potato virus X (PVX) (Giritch et al. 2006; Hiatt and Pauly 2006) each encoding a separate mAb immunoglobulin chain. Specifically, the MCBs for the αCCR5 mAb were produced by engineering the heavy and light antibody chains of αCCR5 into two separate magnICON® provector transfer DNAs (T-DNAs) of *A. tumefaciens* strain GV3101. The particular magnICON system used in this case study requires three additional provector T-DNAs containing integrase, tobacco mosaic virus (TMV) replicase, and potato virus X (PVX) replicase components. The integrase provector produces bacteriophage PhiC31 integrase which catalyzes the site specific fusion of the antibody and replicase provectors when all vectors are delivered together in-planta. All provector T-DNAs were engineered and manipulated in *E. coli* and then individually transformed into electrocompetent *A. tumefaciens* strain GV3101. The MCB and WCBs for αCCR5 consists of five separate transformed Agrobacterium strains. Actually, once the optimization of expression is achieved, there is no need in use of provector system and use of five separate agrobacterial strains. They can be replaced by a more simple strategy where two strains, one carrying assembled PVX-derived, another-TMV-derived viral vectors with light and heavy chains of mAb, respectively.

For production of the αCCR5 mAb, *N. benthamiana* plants were seeded in plant growth trays containing prewetted tobacco mix medium using an automated precision needle seeder. Plants germinated in the nursery for 7 days before being moved to a production greenhouse where they are maintained for inoculation. A dilute Agroinfiltration solution of five WCBs for αCCR5 expression is prepared by combining overnight cultures of each of five MCBs. *N. benthamiana* plants were Agroinfiltrated in commercial scale infiltration chamber biomass and

Table 2 Release test specifications and results for anti-CCR5 mAb

Parameter	Test method	Release specification	Production batch results
Appearance	Visual	Clear, colorless to amber, liquid	Clear, colorless, liquid
Protein concentration	OD_{280}^{a}	0.7–1.3 mg/mL	1.1 mg/mL
Identity[b]	Isoelectric focusing	4–5 Bands pI range 8.4–9.7	5 bands pI 8.4–9.7
Purity	SDS-PAGE[c]	≥95% (Sum of Heavy and Light Chain)	>99%
Purity	Size exclusion HPLC	≥90% Monomer ≤10% aggregation LMW[d] species ≤ 10%	97% Monomer 0.44% aggregation 2.63% LMW
Potency	Viral neutralization	IC50 < 1 μg/mL	0.08 μg/mL
Physical/ Chemical Properties	pH	5.5–6.5 pH units	6.2 pH units
Physical/ Chemical Properties	Conductivity	9.15 mS/cm ± 0.5	9.15 mS/cm
Safety[e]	Endotoxin	<10 EU/mL	0.5 EU/mL
Safety[f]	Bioburden	<10 CFU/mL	<1 CFU/mL
Impurities	1-methyl-2-[3-pyridyl]-pyrrolidine (nicotine) concentration	FIO[g]	<50 ppb
Impurities	Residual host cell protein	FIO	<0.2%

[a] Optical density (*OD*)
[b] Isoelectric focusing polyacrylamide gel electrophoresis (PAGE) gels
[c] Sodium dodecyl sulfate PAGE (*SDS-PAGE*)
[d] Endotoxin units (*EU*)
[e] *LMW* Low molecular weight
[f] Colony forming units (*CFU*)
[g] *FIO* For information only

then transferred into a temperature-controlled growth room for 10 days post inoculation.

Once introduced into the plant, each vector replicates independently and expresses heavy and light chains in the same cells. The two chains self-assemble into authentic and functional mAbs and are secreted to the apoplastic space (Giritch et al. 2006; Gleba et al. 2007, 2008; Gleba and Giritch 2011). The accumulation level of the αCCR5 mAb in plants using the magnICON® system is 250 mg/kg fresh weight transfected plant material at multi-kg scale. These levels are ~10-fold greater than production levels reported for standard transgenic systems (Valdes et al. 2003).

Aerial plant tissues are harvested and plant material subjected to a two-step disintegration process using a pH 8.5 buffer containing anti-oxidants at a rate of

1.0 L buffer per 1.0 kg of plant. After disintegration, the homogenized plant material is fed into a horizontal screw press for initial separation of plant fiber and liquid extract. The plant extract is adjusted with regards to pH < 6.0 and a high purity filter aid was added. The extract is clarified using a plate and frame filter press using 0.3 μm nominal filter pads.

The clarified plant extract is then processed in a GMP compliant clean room environment and directly loaded on to an equilibrated Protein A column. The mAb is eluted via pH shift and conductivity was adjusted for compatibility with ion exchange chromatography. The treated eluant is subsequently loaded on a ceramic mixed-mode ion exchange column for binding. Impurities are removed by washing and the mAb was eluted by sodium chloride gradient and concentrated using tangential flow ultrafiltration against a 30 kDa molecular weight cut off membrane. Bulk drug substance is sterilely filtered (0.2 μm) and dispensed into sealed pyrogen-free 1.0 mL glass vials.

In-process tests and final product involve a number of analytical tests detailed in Table 2. Among the assays employed measure purity and identity assessments using SDS-PAGE and isoelectric focusing followed by Coomassie Brilliant Blue staining and densitometry. αCCR5 mAb potency is determined using a HIV-1 pseudovirus (RHPA4259.7) and MLV (murine leukemia virus) control and the Britelite Plus system in an ELISA format. Values from this potency assay are compared with a CCR5-specific enzyme-linked immunosorbent assays (ELISAs). Due to the use of a Gram-negative bacterial infection to launch magnICON® vectors, the presence of endotoxin is also monitored to ensure freedom from contaminating molecules. Following purification and 0.2 mm filtration, bulk drug substance is vialed as drug product. Final production lots of anti-CCR5 mAb product are subjected to rigorous release testing in accordance with predefined acceptance criteria specifications (Table 2). The product retain highly active mAb product with specific activity measurements within expected values. The purity of the mAb is found to be >99% pure by SDS-PAGE, with 97% of the product in a monomeric form as determined by size exclusion chromatography-HPLC. Non-protein impurities, such as nicotine, are found at very low levels, measured at parts per billion levels (Table 2). The purification process employed for the anti-CCR5 mAb is highly efficacious to remove potentially contaminating endotoxin and bioburden derived from the *Agrobacterium* inoculums or other environmental bacteria. No bioburden is detectable and endotoxin levels were low conforming to USP specifications. The product shows no significant levels of endotoxin (Table 2).

5 Concluding Remarks

magnICON® vectors represent flexible expression vectors, allowing both single chain and dual chain protein products, at levels that are extremely competitive with other commercially used production methods. Combination of such features of the system as high expression level, ability to produce hetero-oligomeric proteins and an efficiency of the *Agrobacterium*-based vector delivery support the superiority of magnICON® over other plant-based production systems. For example, the last feature—dilution of agrobacterial overnight culture to 0.002 $OD_{600,}$ means that for the infiltration of one ton of green plant biomass ca. 200 mL of overnight agrobacterial culture ($OD_{600} = 2.0$) is required. In contrast, for the nonviral vectors (dilution to 0.24–0.64 OD_{600}) requires ca. 24–64 L of overnight culture. The calculation is based on 20% losses (v/w) of *Agrobacterium*-containing infiltration media relative to the fresh weight of green plant biomass. This calculation does not include the fixed volume of infiltration apparatus, necessary to be filled in before infiltration. This might increase the requested volume of Agrobacterial culture 2–10-fold or more, depending on the apparatus design and daily throughput.

Adaptation of Agro-infiltration methods to accommodate large volume of plants using automated approaches facilitates production capacities of kilogram levels of protein product. magnICON® production is also an attractive alternative for pharmaceutical protein production to the current mammalian or microbial cell-based systems. For example, production of 1 kg of recombinant protein using mammalian (CHO cell) culture (expression level 1 g/L; 50% processing yield) would require 2,000 L of cell culture and production trains would require long lead time (ca 1 month), while for the same amount of recombinant protein to be produced using magnICON® system (expression level 1 g/kg FW; 50% processing yield) 2 tons of *N. benthamiana* plants is required. This can be provided by 0.12 ha of greenhouse space (harvest—1.7 kg/m^2). Taking into account that the incubation of infiltrated plants (equivalent to fermentation using CHO cells), this process requires ca. 1 week time and that the plant supply for infiltration can be provided on regular basis at very low cost from spatially separated from the manufacturing site greenhouses. These aspects of manufacturing process are obviously faster and significantly cheaper comparing to the mammalian cells fermentation. We assumed that the downstream processing cost will be relatively similar for both systems. The potential for reduced production costs coupled with the low risk for contamination with human-tropic adventitious agents and other impurities support the use of magnICON® as the next wave for pharmaceutical product production. Recent data (Pogue et al. 2010) demonstrate that magnICON® systems can be adapted to large-scale production of mAbs and other attractive pharmaceutical products.

In addition to the quality of proteins produced from the magnICON® systems, the speed to develop the products is truly impressive. magnICON® can be used to rapidly develop expression vectors and milligrams of product can be produced in as little as 2 weeks and production of grams of product may take only a few weeks more. These timeframes are much shorter than the requirements to transfect, select, establish and characterize mammalian cells, transgenic animal, or traditional plant-based systems. Further, the yields that can be expected from these systems can be quite high, ranging from 0.25 to 0.75 g/kg when extracting >100 kg of crude plant material. These values are ten fold greater than production levels of the same proteins in transgenic plant systems. The example provided in this review shows that the magnICON® can produce correctly folded hetero-oligomeric proteins that show release properties comparable with standard pharmaceutical products attesting to the robustness of plant expression capabilities.

References

Aviezer D, Brill-Almon E et al (2009) A plant-derived recombinant human glucocerebrosidase enzyme—a preclinical and phase I investigation. PLoS One 4(3):e4792

Bendandi M (2008) Aiming at a curative strategy for follicular lymphoma. CA Cancer J Clin 58(5):305–317

Bendandi M, Marillonnet S et al (2010) Rapid, high-yield production in plants of individualized idiotype vaccines for non-Hodgkin's lymphoma. Ann Oncol 21(12):2420–2427

D'Aoust MA, Lavoie PO et al (2009) Transient expression of antibodies in plants using syringe agroinfiltration. Methods Mol Biol 483:41–50

D'Aoust MA, Lavoie PO et al (2008) Influenza virus-like particles produced by transient expression in *Nicotiana benthamiana* induce a protective immune response against a lethal viral challenge in mice. Plant Biotechnol J 6(9):930–940

Dawson WO, Beck DL et al (1986) cDNA cloning of the complete genome of tobacco mosaic virus and production of infectious transcripts. Proc Natl Acad Sci U S A 83(6): 1832–1836

De Muynck B, Navarre C et al (2010) Production of antibodies in plants: status after twenty years. Plant Biotechnol J 8:529–563

Dean M, Carrington M et al (1996) Genetic restriction of HIV-1 infection and progression to AIDS by a deletion allele of the CKR5 structural gene. Hemophilia Growth and Development Study, Multicenter AIDS Cohort Study, Multicenter Hemophilia Cohort Study, San Francisco City Cohort, ALIVE Study. Science 273(5283):1856–1862

Decker EL, Reski R (2008) Current achievements in the production of complex biopharmaceuticals with moss bioreactors. Bioprocess Biosyst Eng 31(1):3–9

Donson J, Kearney CM et al (1991) Systemic expression of a bacterial gene by a tobacco mosaic virus-based vector. Proc Natl Acad Sci U S A 88(16):7204–7208

Fischer R, Vaquero-Martin C et al (1999) Towards molecular farming in the future: transient protein expression in plants. Biotechnol Appl Biochem 30 (Pt 2):113–116

Fischer R, Stoger E et al (2004) Plant-based production of biopharmaceuticals. Curr Opin Plant Biol 7(2):152–158

Gaertner H, Cerini F et al (2008) Highly potent, fully recombinant anti-HIV chemokines: reengineering a low-cost microbicide. Proc Natl Acad Sci U S A 105(46):17706–17711

Gils M, Kandzia R et al (2005) High-yield production of authentic human growth hormone using a plant virus-based expression system. Plant Biotechnol J 3(6):613–620

Giritch A, Marillonnet S et al (2006) Rapid high-yield expression of full-size IgG antibodies in plants coinfected with noncompeting viral vectors. Proc Natl Acad Sci U S A 103(40): 14701–14706

Gleba Y, Giritch A (2011) Plant viral vectors for protein expression. Recent advances in plant virology. Carole Caranta MAA, Mark Tepfer and Lopez-Moya JJ. Caister Academic Press, Norfolk, pp 387–412.

Gleba Y, Marillonnet S et al (2004) Engineering viral expression vectors for plants: the 'full virus' and the 'deconstructed virus' strategies. Curr Opin Plant Biol 7(2):182–188

Gleba Y, Klimyuk V et al (2005) Magnifection–a new platform for expressing recombinant vaccines in plants. Vaccine 23(17–18):2042–2048

Gleba Y, Klimyuk V et al (2007) Viral vectors for the expression of proteins in plants. Curr Opin Biotechnol 18(2):134–141

Gleba Y, Marillonnet S et al (2008) Plant virus vectors (gene expression systems). Enc Virol 3rd edition 4:229–237
Goodin MM, Zaitlin D et al (2008) *Nicotiana benthamiana*: its history and future as a model for plant-pathogen interactions. Mol Plant Microbe Interact 21(8):1015–1026
Goodner B, Hinkle G et al (2001) Genome sequence of the plant pathogen and biotechnology agent *Agrobacterium tumefaciens* C58. Science 294(5550):2323–2328
Hellwig S, Drossard J et al (2004) Plant cell cultures for the production of recombinant proteins. Nat Biotechnol 22(11):1415–1422
Hiatt A, Pauly M (2006) Monoclonal antibodies from plants: a new speed record. Proc Natl Acad Sci U S A 103(40):14645–14646
Hoentjen F, van Bodegraven AA (2009) Safety of anti-tumor necrosis factor therapy in inflammatory bowel disease. World J Gastroenterol 15(17):2067–2073
Horn ME, Woodard SL et al (2004) Plant molecular farming: systems and products. Plant Cell Rep 22(10):711–720
Huang Z, Santi L et al (2006) Rapid, high-level production of hepatitis B core antigen in plant leaf and its immunogenicity in mice. Vaccine 24(14):2506–2513
Huang Z, Phoolcharoen W et al (2009) High-level rapid production of full-size monoclonal antibodies in plants by a single-vector DNA replicon system. Biotechnol Bioeng 106(1):9–17
Jacobson JM, Saag MS et al (2008) Antiviral activity of single-dose PRO 140, a CCR5 monoclonal antibody, in HIV-infected adults. J Infect Dis 198(9):1345–1352
Johansen LK, Carrington JC (2001) Silencing on the spot. Induction and suppression of RNA silencing in the Agrobacterium-mediated transient expression system. Plant Physiol 126(3):930–938
Kearney CM, Donson J et al (1993) Low level of genetic drift in foreign sequences replicating in an RNA virus in plants. Virology 192(1):11–17
Kearney CM, Thomson MJ et al (1999) Genome evolution of tobacco mosaic virus populations during long-term passaging in a diverse range of hosts. Arch Virol 144(8):1513–1526
Klimyuk V, Marillonnet S et al (2005) Production of recombinant proteins in plants. In: Knäblein J (ed) Modern biopharmaceuticals. WILEY-WCH Verlag GmbH & Co. KGaA, Weinheim, pp 893–917
Knäblein J, McCaman M (2003) Modern Biopharmaceuticals-recombinant protein expression in transgenic plants. SCREENING Trends Drug Discov 6:33–35
Ko K, Brodzik R et al (2009) Production of antibodies in plants: approaches and perspectives. Curr Top Microbiol Immunol 332:55–78
Lienard D, Sourrouille C et al (2007) Pharming and transgenic plants. Biotechnol Annu Rev 13:115–147
Lindbo JA (2007a) High-efficiency protein expression in plants from agroinfection-compatible Tobacco mosaic virus expression vectors. BMC Biotechnol 7:52
Lindbo JA (2007b) TRBO: a high-efficiency tobacco mosaic virus RNA-based overexpression vector. Plant Physiol 145(4):1232–1240
Longo D (2009) Lymphoma, Non-Hodgkin's. Encyclopedia of cancer, 2nd ed. Springer, Berlin.
Ma JK, Drake PM et al (2003) The production of recombinant pharmaceutical proteins in plants. Nat Rev Genet 4(10):794–805
Mack G (2008) FDA balks at Myozyme scale-up. Nat Biotechnol 26(6):592
Mallory AC, Parks G et al (2002) The amplicon-plus system for high-level expression of transgenes in plants. Nat Biotechnol 20(6):622–625
Marillonnet S, Giritch A et al (2004) In planta engineering of viral RNA replicons: efficient assembly by recombination of DNA modules delivered by Agrobacterium. Proc Natl Acad Sci U S A 101(18):6852–6857
Marillonnet S, Thoeringer C et al (2005) Systemic *Agrobacterium tumefaciens*-mediated transfection of viral replicons for efficient transient expression in plants. Nat Biotechnol 23(6):718–723
Mett V, Farrance CE et al (2008) Plants as biofactories. Biologicals 36(6):354–358
Moore JP, Trkola A et al (1997) Co-receptors for HIV-1 entry. Curr Opin Immunol 9(4):551–562

Murga JD, Franti M et al (2006) Potent antiviral synergy between monoclonal antibody and small-molecule CCR5 inhibitors of human immunodeficiency virus type 1. Antimicrob Agents Chemother 50(10):3289–3296

Pogue GP, Lindbo JA, Garger SJ, Fitzmaurice WP (2002) Making an ally from an enemy: plant virology and the new agriculture. Ann Rev Phytopathol 40:45–74

Pogue GP, Vojdani F et al (2010) Production of pharmaceutical-grade recombinant aprotinin and a monoclonal antibody product using plant-based transient expression systems. Plant Biotechnol J 8:638–654

Rabindran S, Dawson WO (2001) Assessment of recombinants that arise from the use of a TMV-based transient expression vector. Virology 284(2):182–189

Regnard GL, Halley-Stott RP et al (2010) High level protein expression in plants through the use of a novel autonomously replicating geminivirus shuttle vector. Plant Biotechnol J 8(1):38–46

Rybicki EP (2009) Plant-produced vaccines: promise and reality. Drug Discov Today 14(1–2): 16–24

Sainsbury F, Lomonossoff GP (2008) Extremely high-level and rapid transient protein production in plants without the use of viral replication. Plant Physiol 148(3):1212–1218

Sainsbury F, Lavoie PO et al (2008) Expression of multiple proteins using full-length and deleted versions of cowpea mosaic virus RNA-2. Plant Biotechnol J 6(1):82–92

Sainsbury F, Liu L et al (2009) Cowpea mosaic virus-based systems for the expression of antigens and antibodies in plants. Methods Mol Biol 483:25–39

Santi L, Giritch A et al (2006) Protection conferred by recombinant Yersinia pestis antigens produced by a rapid and highly scalable plant expression system. Proc Natl Acad Sci U S A 103(4):861–866

Santi L, Batchelor L et al (2008) An efficient plant viral expression system generating orally immunogenic Norwalk virus-like particles. Vaccine 26(15):1846–1854

Severinov K, Soushko M et al (1993) Rifampicin region revisited. New rifampicin-resistant and streptolydigin-resistant mutants in the beta subunit of Escherichia coli RNA polymerase. J Biol Chem 268(20):14820–14825

Shearer WT, DeVille JG et al (2006) Susceptibility of pediatric HIV-1 isolates to recombinant CD4-IgG2 (PRO 542) and humanized mAb to the chemokine receptor CCR5 (PRO 140). J Allergy Clin Immunol 118(2):518–521

Stein KE, Webber KO (2001) The regulation of biologic products derived from bioengineered plants. Curr Opin Biotechnol 12(3):308–311

Stoger E, Schillberg S et al (2004) Antibody production in transgenic plants. Methods Mol Biol 248:301–318

Takeda A, Sugiyama K et al (2002) Identification of a novel RNA silencing suppressor, NSs protein of Tomato spotted wilt virus. FEBS Lett 532(1–2):75–79

Trkola A, Ketas TJ et al (2001) Potent, broad-spectrum inhibition of human immunodeficiency virus type 1 by the CCR5 monoclonal antibody PRO 140. Virology 75(2):579–588

Turpen TH, Turpen AM et al (1993) Transfection of whole plants from wounds inoculated with *Agrobacterium tumefaciens* containing cDNA of tobacco mosaic virus. J Virol Methods 42(2–3):227–239

Usha R, Rohll JB et al (1993) Expression of an animal virus antigenic site on the surface of a plant virus particle. Virology 197(1):366–374

Valdes R, Gomez L et al (2003) Large-scale purification of an antibody directed against hepatitis B surface antigen from transgenic tobacco plants. Biochem Biophys Res Commun 308(1):94–100

Van Vloten-Doting L, Bol JF et al (1985) Plant virus-based vectors for gene trans-fer will be of limited use because of the high error frequency during viral RNA synthesis. Plant Mol Biol 4:323–326

Vancanneyt G, Dubald M et al (2009) A case study for plant-made pharmaceuticals comparing different plant expression and production systems. Methods Mol Biol 483:209–221

Vaquero C, Sack M et al (1999) Transient expression of a tumor-specific single-chain fragment and a chimeric antibody in tobacco leaves. Proc Natl Acad Sci U S A 96(20):11128–11133

Vezina LP, Faye L et al (2009) Transient co-expression for fast and high-yield production of antibodies with human-like N-glycans in plants. Plant Biotechnol J 7(5):442–455

Voinnet O, Rivas S et al (2003) An enhanced transient expression system in plants based on suppression of gene silencing by the p19 protein of tomato bushy stunt virus. Plant J 33(5):949–956

Webster DE, Wang L et al (2009) Production and characterization of an orally immunogenic Plasmodium antigen in plants using a virus-based expression system. Plant Biotechnol J 7(9):846–855

Weiner LM, Dhodapkar MV et al (2009) Monoclonal antibodies for cancer immunotherapy. Lancet 373(9668):1033–1040

Weintraub JA, Hilton JF et al (2005) Clinical trial of a plant-derived antibody on recolonization of mutans streptococci. Caries Res 39(3):241–250

Wood DW, Setubal JC et al (2001) The genome of the natural genetic engineer *Agrobacterium tumefaciens* C58. Science 294(5550):2317–2323

Wycoff KL (2005) Secretory IgA antibodies from plants. Curr Pharm Des 11(19):2429–2437

Yang SJ, Carter SA et al (2004) A natural variant of a host RNA-dependent RNA polymerase is associated with increased susceptibility to viruses by *Nicotiana benthamiana*. Proc Natl Acad Sci U S A 101(16):6297–6302

Plant Viral Vectors for Delivery by *Agrobacterium*

Yuri Y. Gleba, Daniel Tusé and Anatoli Giritch

Abstract Plant viral vectors delivered by *Agrobacterium* are the basis of several manufacturing processes that are currently in use for producing a wide range of proteins for multiple applications, including vaccine antigens, antibodies, protein nanoparticles such as virus-like particles (VLPs), and other protein and protein-RNA scaffolds. Viral vectors delivered by agrobacterial T-DNA transfer (magnifection) have also become important tools in research. In recent years, essential advances have been made both in the development of second-generation vectors designed using the 'deconstructed virus' approach, as well as in the development of upstream manufacturing processes that are robust and fully scalable. The strategy relies on *Agrobacterium* as a vector to deliver DNA copies of one or more viral RNA/DNA replicons; the bacteria are delivered into leaves by vacuum infiltration, and the viral machinery takes over from the point of T-DNA transfer to the plant cell nucleus, driving massive RNA and protein production and, if required, cell-to-cell spread of the replicons. Among the most often used viral backbones are those of the RNA viruses Tobacco mosaic virus (TMV), Potato virus X (PVX) and Cowpea mosaic virus (CPMV), and the DNA geminivirus Bean yellow dwarf virus. Prototypes of industrial processes that provide for high yield, rapid scale up and fast manufacturing cycles have been designed, and several GMP-compliant and GMP-certified manufacturing facilities are in place. These efforts have been successful as evidenced by the fact that several antibodies and vaccine antigens produced by magnifection are currently in clinical development.

Contents

Y. Y. Gleba (✉) · D. Tusé · A. Giritch
Nomad Bioscience GmbH, Weinbergweg 22, Halle (Saale), Germany
e-mail: gleba@nomadbioscience.com

Current Topics in Microbiology and Immunology (2013) 375: 155–192
DOI: 10.1007/82_2013_352

Published Online: 15 August 2013

1 Introduction

1.1 Plant Biotechnology and Protein Production

Plant biotechnology is a real albeit still modest industry. Data published by the International Service for the Acquisition of Agri-biotech Applications (ISAAA) show that in 2012, the global area under genetically modified (GM) crops, first introduced in 1996, continued to grow for the sixteenth consecutive year and reached 170 million hectares. Today, 90 % of cotton and 85 % of maize grown in USA, as well as over 50 % of soybeans grown worldwide, are GM. These numbers reflect benefits enjoyed by the various participants in the GM seed business (estimated at over 15 billion US dollars last year). All GM plants grown to date have been modified by incorporation of traits that improve the agronomic performance of crops, in particular by endowing plants with herbicide and insect resistance. GM plants expressing other agronomic traits (modified oil composition, drought resistance, yield increase, etc.) and novel products are also being developed.

The next most advanced class of plant biotechnology products under development today is that of pharmaceutical proteins. Table 1 lists plant-derived pharmaceutical proteins that have been approved by regulatory agencies or are in active clinical-stage development. There are currently three plant-made pharmaceutical proteins approved for commercial sale: An anti-caries antibody (developed by

Table 1 Plant-made pharmaceutical proteins that have been approved or are currently in clinical development

Biopharmaceutical	Approval	Company	Reference
CaroRx™ anti-caries mAb	EU (medical device)	Planet biotechnology	www.planetbiotechnology.com
Newcastle virus subunit vaccine	USA	DowAgroSciences	www.dowagro.com
Taliglucerase™ glucocerebrosidase	USA, Israel, and Brazil	Protalix	www.protalix.com
α-galactosidase A	Phase I–II	Protalix	www.protalix.com
Acetylcholinesterase	Phase I	Protalix	www.protalix.com
Human Interferon α	Phase II	Biolex	www.biolex.com
Human Insulin	Phase II	SemBioSys	www.sembiosys.com
Human lactoferrin	Phase II	Ventria	www.ventria.com
H5N1 Influenza virus VLP pandemic vaccine	Phase I and II	Medicago	www.medicago.com
Influenza virus quadrivalent seasonal vaccine	Phase I	Medicago	www.medicago.com
H1N1 Influenza virus subunit vaccine	Phase I	Fraunhofer USA	www.fraunhofer-cmb.org
Non-Hodgkin's lymphoma individualized vaccine	Phase I	Icon Genetics	www.icongenetics.com
Anti-HIV mAb	Phase I	Pharma-Planta	www.pharma-planta.org

Planet Biotechnology); a Newcastle Disease virus vaccine (Dow AgroSciences); and glucocerebrosidase for enzyme replacement therapy (Protalix Biotherapeutics); the last product being the most significant both in terms of potential market and because of it being the first approved i.v. injectable biopharmaceutical made in plants. It was developed as a biosimilar version of human glucocerebrosidase (the original product being manufactured in animal cells and marketed by Genzyme) by Protalix Biotherapeutics, an Israeli company specializing in the application of cultured plant cells for manufacturing biopharmaceuticals. Protalix successfully completed all necessary clinical trials and won FDA approval to market the product. For clarity, the product was not approved by FDA through the newly established biosimilars pathway but rather through the conventional pathway for novel biologics. Perhaps equally significant, the company also entered into a joint distribution alliance with Pfizer, the world's largest pharmaceutical company, to sell its plant-made therapeutic in the USA, Israel, and Brazil. These events represent notable milestones in the field of plant-made pharmaceuticals, because they demonstrate acceptance of the platform by both a major regulatory agency and a major player in the traditional pharmaceutical industry.

All three products mentioned above were developed through first-generation expression processes (stable nuclear genetic transformation) and their successful approval significantly lowered both real and perceived risks associated with using plants as new manufacturing hosts and thus paved the way for new, more efficient plant-based manufacturing processes. Among those novel processes that have made the most significant progress in recent years are transient expression technologies, especially technologies utilizing plant viral vectors delivered by *Agrobacterium*, a process also called magnifection. The area of plant-made pharmaceuticals is a rapidly developing field and its progress has been the subject of a number of reviews (Gleba et al. 2004, 2005, 2007, 2008; Gleba and Giritch 2011; Canizares et al. 2005a, b; McCormick and Palmer 2008; Lico et al. 2008; Yusibov and Rabindran 2008; Mett et al. 2008; Rybicki 2009; Daniell et al. 2009; Smith et al. 2009; Karg and Kallio 2009; Sainsbury et al. 2009; Komarova et al. 2010; Pogue et al. 2010; Chen et al. 2011; Tusé 2011; Whaley et al. 2011; Mortimer et al. 2012; Pelosi et al. 2012, and in this book, Klimyuk et al. 2012; Saunders and Lomonossoff 2013 etc.).

It is poignant that most of the currently announced or planned clinical trials are based on biopharmaceuticals produced using transient expression systems. Those include recombinant biosimilars, antigens, and immunoglobulins produced using magnifection (Icon Genetics/Nomad Bioscience, Germany; Mapp Biopharmaceutical, USA; Fraunhofer Institute for Molecular Biotechnology (IMB), USA; iBio, USA; Caliber Biotherapeutics, USA; PlantForm, Canada, etc.) or a nonviral transient system (Medicago, Canada). The popularity of transient systems reflects their speed and yields, as well as their scalability, and importantly acceptance by the regulatory agencies. There are several plant-based GMP-compliant or GMP-certified manufacturing facilities that integrate transient expression in their main production processes (Kentucky Bioprocessing, USA; Caliber Biotherapeutics, USA; Medicago, Canada; Icon Genetics, Germany; Fraunhofer IMB, Germany). Despite such commercial interest and diversity, it bears mentioning that plant-made pharmaceuticals is a difficult business area, as most large pharmaceutical companies and investors perceive it as a dual risk business proposition (new product and new process). It is therefore not surprising that the proportion of failures among small companies in this sector is higher than, for example, small companies active in agronomic biotechnology as a whole. Notably, two of the companies listed in Table 1, SemBioSys Genetics (Canada) and Biolex Therapeutics (USA), have ceased operations due to resource limitations.

1.2 Strategies Used in Viral Vector Design

Plant virus vectors have been developed as an alternative to stably transformed transgenic or transplastomic plants (reviewed in Gleba et al. 2004, 2007; Gleba and Giritch 2011) because of some anticipated advantages, including expression speed and yield, reduced cost/duration of research and development, very high throughput, flexibility, and full scalability. In recent years, there has been impressive technical progress in the development of production processes and industrial plant hosts, based on different approaches, including the viral vector-based approach. These studies advanced the technical utilities of plant virus vectors, by designing vectors that are not simply copies of a wild-type virus that carries a heterologous gene sequence, but that, instead, have been improved in various ways, in particular, by delegating rate-limiting functions to agrobacteria or a plant host, thus allowing for a more efficient, versatile, controlled, and safe process. Most progress has been achieved thus far with RNA viral vectors, and the most advanced vectors use only a handful of plant virus species as a platform. Preferred vectors include first of all Tobacco mosaic virus (TMV), but also Potato virus X (PVX), Alfalfa mosaic virus (AMV), and Cowpea mosaic virus (CPMV). Recent data show that vectors based on some DNA viruses such as Bean yellow dwarf virus (BeYDV) can also be successfully used as vectors. Much of the research on viral vectorology is described also in this book.

Wild-type viruses are capable of performing diverse functions, including initial host infection, nucleic acid amplification/replication, protein translation, assembly of mature virions, cell-to-cell spread, long-distance spread, reprogramming of the host biosynthetic processes including shut off of most cell functions, suppression silencing of the host, and other events. Only some of these functions are required, or even desired, in an efficient vector, which, in addition, should be able to perform at least one new function, namely, high-level expression of one or more heterologous genes. Today, there are two different approaches that could be used to develop an efficient viral vector/host process.

Historically, the first approach was to design a vector capable of infecting a plant in the same way as a wild-type virus but that has, in addition, been engineered to carry and express a heterologous sequence coding for a gene of interest (this approach is termed a 'full virus' vector strategy). Progress, in this case, was a result of extensive studies that focused on engineering a vector that was essentially a fully functional virus that, despite carrying and expressing a heterologous sequence, retained infectivity, stability, and systemic virulence in its host. The most extensive program resulting in a series of commercial vectors used in various laboratories worldwide, and many product candidates expressed and tested in preclinical trials, was implemented at Large Scale Biology Corp., Vacaville, CA, USA (LSBC; formerly Biosource Genetics) together with associates from several universities. LSBC also achieved a significant milestone in the year 2000 by having produced the first plant-made vaccines (personalized lymphoma vaccines) that were administered parenterally in a clinical trial under US FDA IND, and in the process established platform feasibility, product safety, and tumor-relevant immunogenicity (McCormick et al. 2008; Tusé 2011). LSBC ceased operations in 2005 and today most of its assets including intellectual property and the first biomanufacturing facility of its kind are owned by Kentucky BioProcesing (KBP, Owensboro, KY, USA). Their efforts focused on TMV as the vector backbone, with the value and utility of the resultant Geneware® viral vector technology having been documented in numerous important papers, including Donson et al. 1991; Turpen et al. 1995; Shivprasad et al. 1999; Turpen 1999; McCormick et al. 1999, 2003, 2008;

Kumagai et al. 1993, 1995, 2000; Rabindran and Dawson 2001; Lindbo 2007a; O'Keefe et al. 2009; McCormick 2011; Tusé 2011, to name but a few.

The other, more recent trend that is the main topic of this review, reflects an ideology that admits inherent limitations of the viral process, and attempts to 'deconstruct' the virus, by eliminating the genes and functions that are limiting (for example, narrow species-specificity, or diversion of plant host resources by the viral machinery to produce its own coat protein) or undesirable (such as ability to create functional infectious virion particles, thus being a potential biohazard), and rebuilding the process by replacing viral functions with analogous functions that are not derived from a virus (a 'deconstructed virus' vector strategy). In its more advanced forms, these 'second generation' expression systems integrate elements of viral machinery, such as RNA/DNA amplification and cell-to-cell movement, along with nonviral processes such as, replicon formation via *Agrobacterium*-mediated delivery of T-DNA encoding a viral vector, or via activation from a plant chromosomal DNA resulting in generation of replicons from a chromosomally encoded pro-replicon or pro-virus. Arguably, the most advanced transient version of this approach is the technology developed at Icon Genetics (Halle, Germany) and trademarked as magnICON® technology and vectors. Icon Genetics and its magnICON® vector technology achieved an important milestone in this field through the creation of a standardized methodology for *Agrobacterium*-mediated gene expression in plants. Various generations of magnICON® vectors are in use by dozens of business and research laboratories worldwide, at application levels ranging from laboratory discovery research to clinical- and commercial-scale manufacturing (a first for said second-generation technology). The technology as well as its applications have been the subject of numerous patents and publications (Marillonnet et al. 2004, 2005; Gleba et al. 2004, 2005, 2007; Gils et al. 2005; Giritch et al. 2006; Werner et al. 2006, 2011; Huang et al. 2006; Santi et al. 2006, 2008; Webster et al. 2009; Vancanneyt et al. 2009; Bendandi et al. 2010, among many others).

2 Vectorology

2.1 The 'Full Virus' Strategy

Under the *'full virus'* scenario, delivery of replicons and initiation of amplification and expression are effected by infecting the host with a mature viral particle or with a full copy of viral DNA/RNA (and recently, also by delivering it as a T-DNA). The vector is essentially a functional virus that, in addition to carrying genes for housekeeping proteins, encodes a protein of interest under a strong viral promoter, such as the Coat Protein (CP) promoter. First, successful results with RNA vectors derived from a TMV have been published by Dawson et al. (1986). Using the same promoter twice (to transcribe the viral Coat Protein and Protein of

Interest genes) is undesirable because it leads to vector instability; therefore the second promoter is usually derived from a phylogenetically related virus. Alternatively, only an essential part of a protein of interest such as an immunogenic epitope is produced as a fusion to the CP. This approach goes back to the works of G. Lomonossoff and colleagues who used CPMV (Usha et al. 1993); later, a monopartite TMV-based vector was proposed for the same purposes by Turpen et al. (1995). The progress in development of new vectors and their use for manufacturing recombinant proteins and nanoscaffolds has been nothing short of impressive and some of it is reflected in other reviews published in this book.

2.2 The 'Deconstructed Virus' Strategy

2.2.1 Delivery by Agrobacterium ('Magnifection')

This strategy reflects a technologist's attempt to part with the inherent limitations of the viral machinery, while keeping the useful viral building blocks in the overall process. Among different functional blocks that can be used outside the integrated viral system are molecular machines that provide for infectivity, amplification/replication, cell-to-cell movement, assembly of virion particles, suppression (shut off) of plant cellular synthesis, RNA silencing suppression, systemic spread, etc. Some of these mentioned machineries are less efficient than others, in particular, a virus' ability to infect the host may be very low and require some mechanical injury or other manipulations to the plant or help from an insect or other organism acting as a vector for efficient virus transmission. Also, a virus's systemic spread process is usually species-specific, can be easily impaired as a result of genetic manipulation, and is often limited to only some tissues or parts of the plant. Conversely, DNA/RNA replication/amplification ability, and subsequent protein expression, are powerful, relatively robust and less species-dependent events that are essential in most applications as 'amplifiers' of the process of interest.

The most advanced version of the '*deconstructed virus*' approach relies on the use of *Agrobacterium tumefaciens* as a bacterial vector that is capable of efficiently introducing DNA copies of the viral vector into plant cells. All necessary events, including processing of an 'agrodelivered' T-DNA to create a DNA or an RNA amplicon, have been shown to occur in plants. 'Agroinfection,' or delivery of infectious viral material as T-DNA-encoded instructions, has been actively used for many years and is in many ways more efficient than whole virus-mediated infection. Agroinfection is definitely much more efficient than using DNA or RNA infectious molecules, and in case of RNA viruses, it is a very inexpensive alternative to in vitro transcription methods used until recently to convert a DNA vector into an infectious RNA molecule. Over last decade, it has been shown that agroinfection can be used to simultaneously start transient gene amplification and high-level expression in all mature leaves of a plant, and that such a transfection route can be inexpensively performed on an industrial scale. This eclectic technology combines the advantages of three biological systems, namely: (1) the speed and expression level/yield of a virus, (2) transfection efficiency and systemic delivery of an *Agrobacterium*, and (3) diverse synthetic capabilities (including posttranslational modifications) along with low production cost in plants. These features enable an industrial production process that does not require genetic modification of plants and is faster, more efficient, better controlled, more flexible, and compatible with the current industrial infrastructure.

2.2.2 Viral RNA Replicons Delivered as DNA Copies

DNA delivery by *Agrobacterium* dates from the early works by the groups of Mary-Dell Chilton (Yadav et al. 1980), Jeff Schell and Marc Van Montagu (Willmitzer et al. 1980), who demonstrated that in the process of 'crown gall' (a plant disease) formation, a DNA portion (T-DNA) of a large TI (tumor-inducing) plasmid of *Agrobacterium tumefaciens* or *A. rhizogenes* is transferred from a bacterium to a plant cell and integrated on a plant chromosome. The process involves multiple complicated steps that are still not fully understood (see, for example reviews by Gelvin (2008, 2009, 2010, 2012); Citovsky et al. (2007); Magori and Citovsky (2011)) but has nevertheless been widely exploited to genetically transform plants since the first publications of tobacco genetic transformation in 1983 (Fraley et al. 1983; Barton et al. 1983). The initial portion of the process (prior to the T-DNA integration on a chromosome) is also being used extensively as a research tool, because DNA encoding a transcriptional unit under plant-specific transcription signals, once it has reached a plant nucleus, is effectively 'transiently' transcribed and translated even in the absence of integration on a plant chromosome, thus providing for a high throughput and rapid method to study expression of different genes and different regulatory genetic elements in a plant organism.

Multiple 'laboratory' versions of the process of 'transient' transfection have been developed much of that work has focused on providing for an efficient contact between a plant cell and a bacterium (Fig. 1). In a most simple version, a bacterial suspension is 'injected' into a plant cell via the blunt end of a syringe (Fig. 1a, b). Since a plant leaf has an extensive intercellular space (up to one-third of the leaf volume) filled mostly with atmospheric gas, such a procedure allows injection of the liquid into an intercellular cavity, thus allowing for the effective access by bacteria to most leaf cells in a blotted spot. On a larger scale, similar result can be obtained through so-called vacuum infiltration (Fig. 1c, d), in which plants are submerged in a bacterial suspension and a mild vacuum (approx. −0.8 to −1.0 bar) is applied for a very short time (30 s to a few minutes); as a result, much of the gas from the intercellular space is removed; upon release of the vacuum, the bacterial suspension is 'sucked' in the intercellular space. Such a process of 'vacuum infiltration' can be performed on an industrial scale inexpensively, and at Kentucky Bioprocessing LLC (Owensboro, KY, USA), a pilot process has been built that can infiltrate up to 1.2 tons of plant biomass per day. Such a process has

Fig. 1 Different ways of delivering agrobacteria into *Nicotiana benthamiana* plant leaves. **a** Infiltration with diluted suspension of *Agrobacterium* cells using the needleless syringe. **c** Infiltration of whole plants with agrobacterial suspension using vacuum. **e** Spraying of plant with diluted agrobacterial suspension containing the surfactant. **b, d, f** Expression of GFP after syringe infiltration **b**, vacuum infiltration **d** and spraying **f** of *Nicotiana benthamiana* leaves with agrobacterial cultures carrying TMV-based viral vectors. Plants were viewed under UV light 7 days post treatment

some limitations; in particular, the plant material has to be grown in trays to enable inversion and submersion of the aerial biomass into the bacterial suspension. However, the existing process is both easily and flexibly scalable and sufficiently inexpensive for producing any high-value (pharmaceutical) protein, and there are a number of products under development that are expected to be manufactured using this process.

Once the contact between a plant cell and a bacterium has been established, the transfer of T-DNA into a plant cell generally occurs very efficiently; therefore, with few exceptions, researchers have not paid much attention to its improvement. An interesting finding of Reavy et al. (2007) illustrates, however, that the process of T-DNA transfer can be significantly improved. In this work, the VirD2 protein of *Agrobacterium*, one of the key elements of DNA transfer, has been shown to be a substrate for a plant caspase-like protease activity in tobacco. Mutagenesis of the VirD2 protein to prevent such a cleavage increased the efficiency of transient gene transfer and expression more than four times in *Nicotiana benthamiana*, petunia, rapeseed, and maize. Given the complexity of the whole process and multiplicity of the bacterial and plant genes involved [see recent reviews of Gelvin (2008, 2009)], there is no doubt that the improvement of *Agrobacterium* as a plant DNA vector has only begun. There is a natural diversity among different *Agrobacterium tumefaciens* strains with regard to their ability to transfer T-DNA efficiently. This diversity has not yet been fully exploited, as most researchers are still using standard laboratory strains originally developed for genetic transformation of plants, such as tobacco and *Arabidopsis*. Research at Nomad Bioscience (Germany) has identified *Agrobacterium* strain CryX, which exhibits unusually high delivery of T-DNA to both *N. benthamiana* and most other dicotyledonous crops tested to date. In *N. benthamiana*, the delivery by CryX is 100–1000 times better than that of other strains, and it reaches the biological limit of 'one *Agrobacterium* cell per one transfected plant cell' (Gleba 2013).

Studies of *Agrobacterium*-based delivery of viral replicons began with seminal works by Grimsley et al. (1986; who coined the term 'agroinfection') and the first paper dealing with RNA cytoplasmic viruses published by Turpen et al. (1993), who demonstrated that TMV could be delivered as a full DNA infectious copy, albeit very inefficiently. The authors showed that over 10^8 bacteria per plant were needed to start one infection process. In our own later work (Marillonnet et al. 2005), we reasoned that the bottleneck of the actual formation of active replicons was the low ability of the primary transcript to leave the nucleus. Since the RNA of cytoplasmic RNA viruses never enters the plant cell nucleus during the viral life cycle, and many viral genomes reveal multiple sequences that are likely to be improperly recognized by the nuclear RNA splicing machinery, an RNA vector that is a simple 'carbon copy' of a wild-type virus is likely to be degraded before it leaves the nucleus. By modifying the DNA copy of a crucifer strain of TMV (crTMV) in various ways, including introduction of silent mutations, changing the codon usage, and adding plant introns, highly active synthetic T-DNA templates for delivery of RNA viral vectors using *Agrobacterium* have been constructed that, when delivered as DNA precursors, provide efficient processing of the DNA information into active amplicons in almost all (>93 %) cells of infiltrated tobacco—a 10^4-fold improvement over nonoptimized DNA templates when delivered to *Nicotiana benthamiana* and a 10^7-fold improvement when delivered to *N. tabacum*. The proposed process required just 10–20 bacterial cells to generate one infection event. The infiltration process also worked with different plant species as hosts. Subsequently, it has been shown that delivery of other viral backbones as DNA copies could be further simplified. For example, we and others found that a DNA carbon copy of a PVX virus does not need to be modified for successful release from the nucleus and similarly, that an unmodified DNA copy of TMV (U1) virus is also processed effectively (Lindbo 2007a, b).

Similarly, multiple improvements have been achieved with vectors derived from DNA viruses. Mor et al. (2003) using BeYDV, a monopartite geminivirus that infects dicotyledonous plants, developed a high-level expression system that utilizes elements of the replication machinery of this single-stranded DNA virus. The replication initiator protein (Rep) mediates release and replication of a replicon from a DNA construct ("LSL vector") that contains an expression cassette for a gene of interest flanked by cis-acting elements of the virus. The authors used tobacco NT1 cells and biolistic delivery of plasmid DNA for evaluation of replication and expression of reporter genes contained within the vector. The mature version of the BeYDV-derived expression vectors (Chen et al. 2011) contain deletions of the viral genes encoding coat and movement proteins (MP) and insertion of an expression cassette for a protein of interest. Delivery of the geminiviral vector to leaf cells via *Agrobacterium*-mediated import produced very high levels of recombinant DNA that can act as a transcription template, yielding high levels of mRNA for the protein of interest. Several vaccine antigens, including Norwalk virus capsid protein and Hepatitis B core antigen, have been expressed using the BeYDV vector at levels up to 1 mg per g of leaf mass. BeYDV replicons can be stacked in the same vector molecule by linking them in tandem, which enables production of multisubunit proteins such as monoclonal antibody (mAb) heavy and light chains.

Recently, Regnard et al. (2010) have designed a new vector based on the mild strain of the geminivirus BeYDV-m that is replicationally released into plant cells from a recombinant *A. tumefaciens* Ti plasmid. pRIC differs from most other geminivirus-based vectors in that the BeYDV replication-associated elements were included in *cis* rather than from a co-transfected plasmid, while the BeYDV capsid protein (CP) and MP genes were replaced by an antigen-encoding transgene expression cassette derived from the nonreplicating *A. tumefaciens* vector, pTRAc. Authors tested vector efficacy in *Nicotiana benthamiana* by comparing transient cytoplasmic expression between pRIC and pTRAc constructs encoding either enhanced green fluorescent protein (EGFP) or the subunit vaccine antigens, human papillomavirus subtype 16 (HPV-16) major CP L1, and human immunodeficiency virus subtype C p24 antigen. The new vector constructs were amplified *in planta* by up to two orders of magnitude by replication, while 50 % more HPV-16 L1 and 3 to 7-fold more EGFP and HIV-1 p24 were expressed.

2.2.3 Industrial Magnifection Technology

Icon Genetics, a German plant biotechnology company and currently a wholly owned subsidiary of Nomad Bioscience, has developed a new-generation expression platform that fully utilizes the advantages of transient vectors and effectively addresses most of the shortcomings of current technologies (Marillonnet et al. 2004, 2005; Gils et al. 2005; Giritch et al. 2006; Santi et al. 2006; Gleba et al. 2004, 2005, 2007; Gleba and Giritch 2011; Webster et al. 2009; Bendandi et al. 2010; Werner et al. 2006, 2011). The 'magnifection' process (a term derived from 'magnum infection,' an indication that two infection agents are being used, a bacterium and a virus) trademarked as magnICON® and covered by numerous patents/patent applications is a simple and industrially scalable protocol for heterologous protein expression in plants; it is devoid of stable genetic transformation of a plant and relies instead on transient amplification of viral vectors delivered as DNA precursors by *Agrobacterium* to multiple areas of a plant body (systemic delivery). Such a process is in essence an infiltration of whole mature plants with a diluted suspension of bacteria carrying T-DNAs encoding viral RNA replicons. In this process, the bacteria assume the (formerly viral) functions of primary infection and systemic movement, whereas the viral vector provides for cell-to-cell (short distance) spread, amplification, and high-level protein expression. A few adult tobacco plants are sufficient for early construct optimization and fast production of milligram to gram quantities of recombinant protein for preclinical or clinical evaluation, or, in case of individualized vaccines, even for manufacturing. The industrial scale version is essentially the same, but it is built around fully assembled viral vectors and requires simple apparatuses for high-throughput *Agrobacterium* delivery to whole plants.

The magnifection platform effectively addresses most of the major shortcomings of earlier plant-based production technologies by providing an overall best combination of key functional features, including the highest expression levels (approaching biological limits), highest relative yields, ability to express complex proteins, low up- and downstream costs, very fast and low cost R&D, low cost of goods, unparalleled flexibility, and low biosafety risks. The process has been brought to GMP compliance at Icon's new pilot manufacturing facility in Halle, Germany, and in 2009 further scaled for manufacturing by Kentucky BioProcessing (USA) to a throughput capacity of 1.2 tons/day based on Icon's technology. There are several magnICON®-based products, mainly pharmaceutical proteins that are being developed by Icon and its licensees; Icon's lead product candidate, individualized vaccines for Non-Hodgkin's lymphoma, were produced in a cGMP-certified manufacturing facility in Halle and successfully completed clinical trials in 2013 under US FDA IND (Table 1).

Depending on the vector used, the host organism and the initial density of bacteria, this much more efficient and synchronous agro-mediated systemic transfection leads in 4–10 days to a dramatic synthesis of proteins of a wide range of sizes and complexity, providing yields as high as 5 g of product per kg of fresh leaf biomass. The viral vectors effectively shut down normal biosynthetic processes in cells. As a result, in 10–14 days post infection, the most abundant plant leaf protein, ribulose-1,4-bisphosphate carboxylase-oxygenase (RUBISCO), all but disappears or its level is severely reduced, and the amount of recombinant protein reaches as much as 80 % of the total soluble protein in a leaf. Furthermore, since the viral vector is devoid of the CP gene because there is no need for infectivity or systemic movement, it can express longer genes (up to 2.3 kb long inserts or up to MW 80 kDa proteins), and no synthetic potential or resources are wasted on production of the viral CP. Infiltration of plants or detached leaves with bacteria has been done in many different ways, the one simple process being vacuum infiltration by immersing whole aerial parts of plants in a bacterial suspension and applying a weak vacuum (approx. −0.8–1.0 bar) for 10–30 s. Further development and automation of these protocols could increase the throughput of the process and reduce the amount of required manpower (Klimyuk et al. 2012).

2.2.4 Expression Levels Achievable Using Viral Vectors

Much of the efforts in vector and process design, especially those devoted to manufacturing recombinant proteins in plants, have aimed to maximize the recombinant protein yield. Summarizing the data presented in the literature is difficult because different authors have used different measurement approaches. In a plant such as *Nicotiana benthamiana*, a kg of fresh leaf biomass contains approximately 6–8 g soluble protein, and ribulose-bisphosphate carboxylase (RUBISCO) accounts for approximately 50 % of that amount. The capacity of the expression system and the viral vector is usually measured by the amount of a reporter protein produced that is neutral to the plant and easy to express, such as green fluorescent protein (GFP), and measuring the output as an absolute yield (in grams recombinant protein per kg of fresh leaf/green biomass) or as a relative yield (percent of recombinant protein in the total soluble protein). Deconstructed vectors that have no systemic movement capacity are infecting only leaves, whereas stems and petioles (and vascular tissues of leaves) remain uninfected; therefore, it is appropriate to measure both the yield for the infected organs as well as total yield per kg of all fresh green biomass. Full viral vectors, on the other hand, do not infect all aerial (harvestable) parts of the plant and some of the biosynthetic potential is spent on expression of the viral CP; here again, one aims at measuring the yield in fully infected parts as well as averaged yield per kg of green biomass. Since many known viral vectors such as TMV often shut off general plant biosynthesis and in the process of infection the amount of host protein, especially that of RUBISCO, is dramatically reduced, relative measurements such as percentage of total soluble leaf protein could be misleading. Finally, many published data represent measurements based on local transfection of small blotted spots of leaves with the transfection method being not directly scalable; thus, the results are often exaggerated and not always indicative of the potential industrial yields.

Nevertheless, different approaches and use of at least two different viral vector types (TMV- and PVX-based) indicate that transfection usually results in massive accumulation of the recombinant protein, reaching direct yields of up to 5 g of recombinant protein (or recombinant protein plus viral coat protein, in the case of full viral vectors) per kg of infected biomass, or 80 % or more of the total soluble protein. Since it is safe to assume that plant cannot quickly accumulate significantly more total protein than it already had before the infection, these reported yields clearly demonstrated that TMV- and PVX-based vectors provide yields approaching the biological limits of plant-based production. The potential of many alternative vector systems utilizing other viral backbones has yet to be brought to the levels of the established systems utilizing TMV and PVX. In other cases, the expression product is not an individual recombinant protein but a virus or a virus-like particle. Available data, again, show that in many cases the expression levels are high and approach the biological limits of the plant systems.

Table 2 provides yield data for selected recombinant proteins of practical interest based on publications by numerous groups as well as our unpublished results. It is obvious that many proteins, including multiple biopharmaceuticals (interferons, insulins, growth hormones, colony stimulating factors, numerous antigens, selected full IgG antibodies, etc.), can be expressed at very high yields of 1–5 g/kg. Of course, there are also proteins that do not express well, and reasons behind such low expression are sometimes obvious and not connected with the expression method or the plant as a production host. For example, our own attempts to improve expression of proteases, such as trypsin, have reached levels of 0.3 g/kg, and further increase of the protease in an active form is unlikely because of very high toxicity of the enzyme to plant cells. In other cases, such as viral antigens, the reasons behind the relatively low expression levels of full molecules are less obvious; however, many of those molecules do not express well in other expression systems either. In conclusion, available viral expression systems represent mature technical processes that are fully competitive, and for many commercially interesting molecules, viral vectors provide the best expression yields known in the literature (Table 2).

2.2.5 Expression of Complex Proteins

It is possible to direct the foreign protein to various subcellular locations including the endomembrane system, the cytosol, organelles, or even to assembling virion particles. Careful study and control of subcellular targeting of foreign proteins or peptides allows us to maximize their yield, specific activity, stability, and even facilitate their subsequent downstream processing (Marillonnet et al. 2004). Most proteins expressed to date using magnifection have been properly processed, folded, and functional. This is true for high molecular weight single-polypeptide proteins (up to 120 kDa), proteins requiring specific cleavage of signal or transit peptides, extensive disulphide bond formation, complex processing (for example hemagglutinin of Influenza virus), and proteins forming virus-like particles (VLPs) (such as Hepatitis B core protein, Norwalk virus capsid protein, and Papilloma virus L1 protein) or protein and protein/RNA scaffolds (viral particles with fusions

Table 2 Examples of high expression levels of recombinant proteins produced by magnifection in plants

Protein	Vector backbone	Yield, mg/gfw	Reference
GFP, DsRed	TMV	5.0	Marillonnet et al. 2004
GFP	TMV	2.5–4.0	Marillonnet et al. 2005
Human somatotropin	TMV	1.0	Gils et al. 2005
IgG1, mAb A5	TMV/PVX	0.5	Giritch et al. 2006
TMV CP-Protein A	TMV	2–3	Werner et al. 2006
Yersinia pestis F1, V, F1-V	TMV	1–2	Santi et al. 2006
Hepatitis B core antigen	TMV	2.4	Huang et al. 2006
GFP	TMV	5.5	Lindbo 2007a, b
AT adenosine kinase	TMV	5.0	Lindbo 2007b
Tuberculosis ESAT6 antigen	TMV	0.8	Dorokhov et al. 2007
HPV antigen L1	TMV	0.5	Maclean et al. 2007
Plasmodium PyMSP4/5 antigen	TMV	2.0	Webster et al. 2009
Norwalk virus CP	TMV	0.8	Santi et al. 2008
GFP, DsRED	CPMV*	1.6	Sainsbury and Lomonossoff 2008
Anti-HIV IgG	CPMV*	0.33	Sainsbury and Lomonossoff 2008
Hepatitis B core antigen	CPMV*	1.0	Sainsbury and Lomonossoff 2008
Hepatitis B core antigen	BeYDV	0.8	Huang et al. 2009
Norwalk virus CP antigen	BeYDV	0.3	Huang et al. 2009
Anti-Ebola IgG	BeYDV	0.5	Huang et al. 2010
Griffithsin	TMV	1.0	O'Keefe et al. 2009
Aprotinin	TMV	0.3–1.0	Vancanneyt et al. 2009
Non-Hodgkin's lymphoma, over 40 different IgG	TMV/PVX	0.2–4.8	Icon Genetics data
Thaumatin	TMV	2.0	Icon genetics data
scFv, various	TMV	1.0	Icon genetics data
Mini-Insulin	TMV	5.0	Nomad bioscience data
Human Interferon α	TMV	5.0	Nomad bioscience data
Griffithsin	TMV	2.5	Nomad bioscience data

TMV tobacco mosaic virus; *PVX* potato virus X; *BeYDW* bean yellow dwarf virus; *CPMV* cowpea mosaic virus; *CPMV** nonreplicating portions of RNA-2; expression levels are in mg protein per gram fresh leaf biomass

of various domains to its coat proteins). Importantly, extensive experimental data demonstrate that proteins expressed using transient systems are functional. In particular, enzymes and enzyme inhibitors expressed in these systems turned out to be fully active, antigens tested in animal studies were properly immunogenic, and plant-made antibodies were capable of recognizing their respective antigens (Table 3).

Table 3 Examples of subunit antigens and virus-like particles made in plants, with indication of the vector used for expression and the animal models used to test the immune responses

Antigen	Vector	Immune response	Reference
Yersinia pestis F1, V, F1-V	TMV	guinea pig	Santi et al. 2006
Hepatitis B core antigen	TMV	mouse	Huang et al. 2006
Rabbit papillomavirus L1	TMV	rabbit	Kohl et al. 2006
Bacillus anthracis	TMV	mouse	Chichester et al. 2007
Yersinia pestis F1, LcrV	TMV	macaque	Mett et al. 2007
Human papillomavirus L1	TMV	mouse	Maclean et al. 2007
Smallpox virus B5	TMV	Mouse, minipig	Golovkin et al. 2007
Human papillomavirus E7	TMV	Mouse	Massa et al. 2007
HIV-1 Pr55Gag, p24	TMV	Mouse	Meyers et al. 2008
Norwalk virus VLP	TMV	Mouse	Santi et al. 2008
Influenza virus H3N2 HA	TMV	Mouse	Shoji et al. 2008
Influenza virus H1N1 HA	TMV	Mouse	Shoji et al. 2013
Plasmodium PyMSP4/5	TMV	Mouse	Webster et al. 2009
Norwalk virus CP	TMV	Mouse	Santi et al. 2008
Non-Hodgkin's lymphoma murine A20 antigen	TMV/PVX	Mouse	Bendandi et al. 2010
Cholera toxin antigen B	TMV	Mouse	Hamorsky et al. 2013b
B. anthracis protective antigen	TMV	Rabbit	Chichester et al. 2013

TMV tobacco mosaic virus; *PVX* potato virus X

2.2.6 Co-Expression of Two or More Proteins or Polypeptides

Production of antibodies represents a practical case of the need for expression of more than one polypeptide in the same cell. The ability of plants to express full-size human antibodies was discovered over 20 years ago (Hiatt et al. 1989; Ma et al. 1995); however, existing protocols at the time suffered from various shortcomings. Stably transformed plants expressed and correctly folded functional antibodies of IgG and IgA classes, but yields were generally very low (1–25 mg/kg of plant biomass), and, because of the long time necessary to generate transgenic plants, the first gram of research antibody material required more than 2 years to produce. Transient systems, on the other hand, allow production of research quantities of material much faster, but the early versions of transfection systems, such as *Agrobacterium*-mediated transient expression using transcriptional vectors (usually based on 35S CaMV promoter) or first-generation viral vector-mediated expression, have provided for low co-expression of two or several polypeptides necessary for the assembly of heterooligomeric proteins, in particular IgG antibodies (Verch et al. 1998; Vaquero et al. 1999). Subsequent attempts to express two or more different polypeptides from the same viral replicon have largely failed, because of drastic reduction in expression when the same vector was used to express two or more polypeptides.

Recently, however, several rapid and fully scalable processes for high-level expression of functional full-size monoclonal antibodies of IgG class in plants have been developed. Our own process relied on synchronous co-infection and co-replication of two viral vectors, each expressing one antibody chain. The two vectors in question are derived from different noncompeting plant viruses, and unlike two vectors derived from the same virus, they effectively co-express heavy and light chains in the same cell throughout the plant body, resulting in some cases in very high yields (up to 4.8 g/kg fresh biomass) of assembled IgG molecules. These vectors are built from the backbones of two different noncompeting viruses, TMV, and PVX. The protocol was shown to work with many different mAbs of IgG1 and IgG2 classes; the molecules were fully functional and the first grams of material were produced in less than 2 weeks (Giritch et al. 2006; Hiatt and Pauly 2006; Bendandi et al. 2010; Whaley et al. 2011).

Unlike the conventional two-vector-based approach, Hamorsky et al. (2013a) recently designed a system to overexpress full-length IgG1 from a single polypeptide by means of kex2p-like enzyme recognition sites introduced between the heavy and light chains. An enzyme-linked immunosorbent assay (ELISA) revealed that gp120-binding VRC01 IgG1 was maximally accumulated 5–7 days following vector inoculation, yielding $\sim$150 mg of the bnMAb per kg of fresh leaf material. Another alternative approach has been developed by Huang et al. (2010) (see also Chen et al. 2011) who have demonstrated that a BeYDV-based (DNA virus) vector allows expression of antibodies at levels reaching 0.5 mg/g. Other authors (Sainsbury and Lomonossoff 2008; Sainsbury et al. 2010) have used replicative and nonreplicative systems (involving a constitutive 35S promoter and positioning the gene(s) of interest between the 5′ leader sequence and the 3′ untranslated region derived from RNA-2 of CPMV for improved stability); the yield of antibodies was as high as 0.3 mg/g. Collectively, these studies show that complex molecules such as IgG antibodies can be produced in plants using different transient processes; however, so far, the viral replicon-based process still provides for the highest expression and it is expected to be more easily scalable.

Expression of more than two polypeptides is even more of a challenge, especially for traditional TMV- and PVX-derived vectors. Thuenemann et al. (2013) demonstrated recently that Bluetongue VLPs requiring co-expression of four different components can be produced and assembled in *Nicotiana benthamiana* using the CPMV-based HyperTrans (CPMV-HT) and associated pEAQ plant transient expression vector system. The VLPs were shown to elicit a strong antibody response in sheep, and furthermore they provided protective immunity against a challenge with a South African BTV-8 field isolate. The results show that choice of appropriate viral vectors can enable production of immunologically relevant, complex, and heteromultimeric structures that require the co-expression and correct assembly of multiple components.

2.2.7 Plant Viruses as Nanoscaffolds for Novel Biomaterials

Due to their relatively simple macromolecular organization and very high accumulation titers, plant viruses provide an extremely inexpensive source of biopolymers of discrete nanoparticle sizes and configurations than can be manufactured rapidly and under very simple conditions (reviewed by Smith et al. 2009). For example, TMV, the best known plant virus, is also one of the most extensively studied viruses regarding its particulate structure. Its RNA genome is a single 6.4 kb RNA molecule that encodes four proteins, including the 17.5 kDa Coat Protein, the most abundant viral product and the only component of the TMV capsid. Over 2,100 copies of CP fully protect the single-stranded viral RNA, resulting in a rigid rod-shaped viral particle with a length of 300 nm, a diameter of 18 nm, and a molecular mass of 40,000 kDa. The genomic RNA is packaged inside of a 2 nm wide canal formed by the assembled CP capsid. Ninety-five percent of the mass of TMV particles consists of the CP. TMV accumulates to levels of up to 7 g per kilogram of leaf biomass, and therefore the CP represents the most abundant individual protein that can be harvested from plants. Other plant viruses used in nanobiotechnology research include the extensively studied CPMV (Evans 2009), Cowpea chlorotic mottle virus, and several others. The virions can be purified industrially using simple and inexpensive 'low-tech' protocols. Due to the ability of the CP to polymerize in vivo and in vitro and the high stability and defined size of the assembled virions, the CP and the whole viral particle represent a potentially promising biopolymer feedstock for a number of applications in nanobiotechnology (Steinmetz et al. 2006; Young et al. 2008; Evans 2009).

The CP-based polymer can be modified in different ways. One approach consists of adding novel moieties to viral particles in vitro by chemical modification. This can be done by exploiting existing reactive sites of natural viruses and using a variety of reactions such as diazonium coupling/oxime formation or attaching amines to glutamic acid side chains, or using metal-binding sites present on many viral surfaces (reviewed in Evans 2009). Alternatively, simple mutations can be introduced into the CP gene resulting in addition to the virus surface of reactive amino acids such as cysteine or lysine; those can later be used for chemical coupling of multiple components (e.g., Royston et al. 2008).

The other avenue consists of genetically reprogramming the virus to express protein fusions thus adding new functional blocks to the viral core monomer, the CP. Since the structure of many viral particles has been determined at atomic resolution, fusion proteins can be designed in such a way that the added sequences are predicted to lie at the surface of the assembled viral particle (e.g., Werner et al. 2006).

This area of virus-based nanobiotechnology is still in its infancy and far behind similar research done with bacteriophages, but plant viruses will likely offer more practical solutions in real industrial processes. A more detailed review of the nanobiotechnology opportunities based on plant viruses as designer blocks is reviewed in multiple works, including Young et al. (2008); McCormick and Palmer (2008); Smith et al. (2009); Evans (2009); Saunders and Lomonossoff (2013), to name a few.

2.2.8 Spraying with Agrobacteria and Expression of Agronomic Traits

It should be mentioned that when the efficiency of the process is not essential (for example, in the delivery of 'full virus' vectors capable of systemic movement, or in screening rare events such as genetic transformants), bacteria can also be delivered by spraying plants in the presence of a surfactant and/or abrasive (Azhakanandam et al. 2007) or by 'agrodrenching' plants (pouring bacterial solution into soil below the plants (Ryu et al. 2004; Yang et al. 2008). The mentioned publications described the application of full viruses, and hence the efficiency of bacterial delivery per se was impossible to quantify.

Recently, research at Nomad Bioscience GmbH has demonstrated that green plants can be transfected by simply spraying them with agrobacteria in combination with surfactants and/or abrasives (Fig. 1e, f). Between 0.2 and 2 % of leaf cells in the resultant plants can be routinely transfected and they express DNA delivered by a bacterial vector. Based on these experiments, a new biotechnology process has been proposed that takes advantage of this phenomenon and uses initially transfected leaf cells as launching pads to generate secondary molecules capable of moving from initial cells to other cells and organs of a plant, and of transiently reprogramming a plant in numerous practical and useful ways. The secondary molecules currently under evaluation at Nomad include transcription factors, small regulatory and interfering RNAs, and small molecules such as hormones. At least theoretically, they allow us to address an almost infinite number of input and output traits including yield, resistance to abiotic stresses, pathogen resistance (insect, nematode, bacterial, viral), time to flowering, and modulation of protein/oil/carbohydrate levels and quality. We have successfully proved the principle with a number of traits including insect and bacterial resistance, drought tolerance, control of flowering, and fruit set time. The process works with all economically important dicot crops, and we also have evidence that the process works with many commercially important monocots.

Hence, the proposed transient reprogramming of plants is an alternative to genetic transformation of plants and offers a number of potential advantages (the genetic material is not transmitted to seed; the business based on this technology does not require ownership of plant germplasm; the R&D process and time to market are much faster and flexible; the treatments can be performed many times over and multiple traits can be bundled without physical stacking; the technology can be rapidly adapted to reflect rapidly evolving pathogens, etc.). It is also an alternative to using pesticides and agrochemicals as it theoretically allows a much more sophisticated way of protecting plants with lower potential environmental concerns.

When commercially exploited, Nomad's 'agrospray' approach to economically modulate agronomic traits and manufacture commodity materials in the field has the potential to become a disruptive breakthrough advancement in plant biotechnology (Giritch et al. 2012; Hahn et al. 2013). Presently, the technology is at an early stage of development and carries two kinds of risks, namely, technical and regulatory. Both risks are being addressed by Nomad researchers in a systematic and clearly defined way. In 2012, Nomad entered into a long-term strategic collaboration with Bayer CropScience (BCS; Monheim, Germany) that is mutually beneficial and allows BCS rapid access to this technology in a broad and essentially de-risked path. Certain trait categories and crops have been specifically excluded from the Bayer alliance, and Nomad intends to develop those areas with other strategic partners.

The agrospray approach also opens up new possibilities in biomaterial manufacturing, especially in cases of low-cost high-volume products. If simple spraying of *Nicotiana* plants with *Agrobacterium* vectors in the presence of a surfactant is combined with agrobacterial T-DNA encoding viral replicons capable of cell-to-cell movement, up to 90 % of the leaf cells can be transfected and express a protein of interest at levels up to 50 % of total soluble protein. This simple, fast, and indefinitely scalable process was successfully applied to produce cellulases, one of the most high-volume and cost-sensitive biotechnology products. Hahn et al. (2013) demonstrated for the first time that representatives of all hydrolase classes necessary for cellulosic biomass decomposition can be expressed at high levels, stored as silage for extended periods without additional processing or significant loss of activity, and the silage used directly as enzyme additives. This process enables production of cellulases at commodity agricultural prices and could find broad applicability in the large-scale production of many other cost-sensitive proteins.

2.2.9 Magnifection Vectors as a Research Tool

A number of methods relying on different viruses are being used to express proteins or RNAs in plants for research purposes. These methods rely on different modifications of the core viral design. Since the general purpose is rapid and high-throughput expression, and only small (usually submilligram) amounts of protein are usually required, many existing research vectors leave out the viral component(s) providing systemic movement.

Use of DNA agrodelivery for starting an RNA-based amplification provides new opportunities for vector engineering, since the transiently delivered DNA in a plant cell can be effectively manipulated prior to conversion into a desired RNA amplicon. In particular, it has been shown that multiple T-DNAs transiently co-delivered from one or more agrobacteria can be efficiently recombined by site-specific recombinases, such as Cre recombinase or integrase from phage φc31. The transcribed RNA can be cleaned of the resultant remaining recombination sites by engineering them to be within an intron that is afterward cleaved by the nuclear machinery. The resultant process allows for simple and inexpensive engineering of RNA vectors in vivo and assembly of the expression cassettes, through simple mixing and co-infiltration of plants with agrobacteria carrying different components of the vectors as T-DNAs, as well as an integrase (Marillonnet et al. 2004). In particular, rapid expression and assembly of proteins by adding different targeting or signal peptides, adding binding domains, fusion proteins, purification domains or cleavage sites, can all be achieved by simply mixing prefabricated strains of bacteria each harboring desired vector modules, and can provide milligram quantities of proteins expressed in different plant compartments or as different fusions, within a few days.

The recent development of rapid DNA manipulation 'toolboxes' of such as 'Golden Gate' cloning, gene shuffling and vector assembly technology that relies on IIs restriction endonucleases, is likely to improve the speed of construct engineering even more, and when coupled with simple viral-based expression assays, can provide for the ultimate in high-throughput screening of gene candidates in plants (Engler et al. 2008, 2009; Weber et al. 2011; Thieme et al. 2011).

2.3 Inducible Release of Viral Replicons From a Chromosome of a Stably Transformed Plant

Alternative possibilities, explored by different research groups, have been to introduce the viral amplicon into a plant and integrate it as a DNA locus on a chromosome with the goal of combining the RNA amplification and protein expression power of viral machinery with the simplicity of handling a transgenic plant in the field. The replicon is engineered to be inducible, in order to allow early accumulation of plant biomass (growth phase) and, upon induction, to provide the most efficient conversion of plant materials and energy into recombinant protein (production phase). In such a case, the speed of product development slows, as the transformation and generation of a transgenic plant requires additional months of time, and the development of an industrial production host usually takes many years. This is, however, not always a major limiting factor in many product development programs. In the current prototypes of such a process, the DNA locus in question is not per se sufficient to trigger active amplification, because the amplicon is encrypted and requires subsequent activation, for example through site-specific recombination, or through suppression or due to early silencing by the plant host. Such activation can later be achieved, for example, through induction of a recombinase that may be present in the transgenic plant and is controlled by a chemically inducible promoter. The first example of this approach has been proposed by Mallory et al. (2002) who demonstrated that a PVX-based amplicon released from a transgenic plant is usually silenced, but it can be re-activated by hybridization with a plant that expresses a suppressor of posttranscriptional gene silencing (PTGS), such as the proteinase HcPro from Tobacco etch virus. Starting with the papers of Mallory et al. (2002) and Voinnet et al. (2003), the HcPro or P19 PTGS suppressors are being widely used today in different transgenic and transfection processes.

Integration of the viral vector into a plant host's chromosome offers multiple technical solutions for overcoming the strict limitations of classical viral vectors. By activating the amplicon from a chromosome, the process is being initiated independently in multiple cells throughout the whole plant; thus the slow, usually incomplete and highly species-specific process of systemic spread is eliminated. The resultant cascade is also much safer, as it can occur only in a GM organism, and it does not yield mature viral particle capable of secondary infection.

Several research groups have published various versions of inducible viral systems. Dohi et al. (2006) have proposed an inducible system that relies on an estrogen-inducible promoter in BY2 cultured tobacco cells. Upon induction with estrogen, a modified TMV vector expressing GFP instead of the CP was shown to amplify and express the gene of interest at levels of up to 10 % of total soluble protein. Neither the viral RNA nor GFP were detectable in noninduced cells. It is difficult to rate the efficiency of this system because the expression levels of GFP in simpler viral systems is significantly higher. Zhang and Mason (2006) developed an inducible system based on the geminiviral Bean yellow dwarf virus. Release of replicon from a stably inserted construct was obtained by expression of the replication initiator protein in *trans* that has been placed under control of the ethanol-inducible promoter. Transgenic plants were shown to induce release of the vector and episomal replication to high copy number after ethanol treatment, resulting in up to an 80-fold increase of mRNA levels and up to a 10-fold increase of the translation product. Tremblay et al. (2007) developed an inactivated DNA replicon of Turnip mosaic virus (TuMV), which was reactivated by a recombination event based on the Cre-loxP system; viral replication has been prevented by the insertion of a translation terminator sequence flanked by two loxP sites at the junction of the P1-HCPro-coding genes and removal of the translation terminator sequence was achieved by the controlled expression of Cre. Delivery of the Cre recombinase to *Arabidopsis* plants transgenic for the inactivated TuMV replicon was obtained by agroinfiltration, PVX-based production, or transgenic chemical-inducible expression. In each case, reactivation of TuMV replication was observed. In a different approach, Hull et al. (2005) used viral vector delivered by agroinfiltration as a transgene expression switch, by using a TMV-based vector to deliver GAL4-VP16 transcriptional activator. The system displayed tight control in both transient and transgenic assays, but so far, there are no results further describing this system and its use.

While technically interesting, the studies mentioned above did not provide broadly applicable, useful solutions for plant-based manufacturing, one important reason being the lack of sufficient control. An important and essential advance was the design of an effective switch that provided tight control and could be effectively used in industrial processes. Such systems would require very stringent control and sufficient induction multiples. In this regard, we have to borrow more from the design of similar expression systems developed for bacterial, fungal, plant and mammalian cells [reviewed in Gatz (1997); Padidam (2003); Terpe (2006); Weber and Fussenegger (2007); Corrado and Karali (2009)]. One specific requirement for such chemically inducible systems in industrial settings is the use of small-molecule chemical inducers that are approved for field applications, commercially available and safe. Candidates fulfilling these criteria include ethanol, used as a part of alcohol-inducible system (Caddick et al. 1998), and the commercial insecticide methoxyfenozide, acting as an agonist for the ecdysone receptor (Zuo et al. 2000; Padidam et al. 2003).

A relatively mature and industrially relevant inducible system for plants has been recently described by Werner et al. (2011), consisting of a novel ethanol-inducible expression process that provides efficient release of viral RNA replicons from DNA pro-replicons contained in a stably transformed cassette. To achieve tight control of replicon activation in the noninduced state, the viral vector has been deconstructed, and its two components, the replicon and the MP, have each been placed separately under the control of an inducible promoter. Transgenic *Nicotiana benthamiana* plants incorporating this double-induction system demonstrated negligible background expression, high (over 0.5×10^4-fold) induction multiples, and high absolute inducible levels of protein expression (up to 4.3 mg recombinant protein per g of fresh green biomass) approaching closely the levels achievable with transient expression systems (Marillonnet et al. 2005), and thus the biological limits of recombinant protein expression in plants. The protocol is indefinitely scalable, supports expression of important recombinant proteins, and as such can be directly used for industrial manufacturing.

3 Plant Manufacturing Hosts, Post-Translational Modifications, and Plant-Specific Sugars

3.1 The Importance of Plant Glycosylation

With a few exceptions (insulin, serum albumin, apolipoprotein A), proteins made by the human body and consequently those used as therapeutics are glycosylated. Glycans on the protein surface are sometimes not essential and do not influence clinical performance; examples of such cases include some interferons and colony-stimulating factors. In a few other cases, such as erythropoietin, a complex and specific glycosylation is critical for therapeutic efficacy. For many other therapeutic proteins, correct glycosylation has proven an essential prerequisite to obtaining desired utility. Research at different laboratories demonstrated that the glycan challenge can be reliably solved through engineering plants that express altered, or if necessary, humanized glycans on the surface of biopharmaceuticals.

Plants are capable of assembling complex proteins; in cellular compartments such as the Golgi, different proteins of human origin are properly processed, cleaved, folded, assembled, and posttranslationally modified. The only recognized essential difference in posttranslational modification by a plant cell relative to mammals is glycosylation. Although plants add the main sugar scaffolds to the protein, the resultant glycoforms are different from those made by the human body in several important ways. Unlike human cells, plant cells add D-xylose β-1,2-linked to the core β-linked mannose and L-fucose α-1,3-linked to the first core *N*-acetylglucosamine (GlcNAc) residue of the glycan. Human cells lack xylose altogether, and add fucose but through an α-1,6-linkage. These sugars are expected to be weakly immunogenic and are probably not desirable in pharmaceutical proteins intended for intravenous applications. Plants mainly add complex type, diantennary *N*-glycans terminating in mannose or *N*-acetylglucosamine, and importantly plant-made glycoproteins lack galactose and *N*-acetylneuraminic acid (sialic acid) provided by human cells.

A special case worth mentioning is plant-based production of recombinant antibodies, in particular, because many attempts are being made to produce plant-based 'biosimilar' or 'biobetter' versions of marketed antineoplastic antibodies such as Avastin®/bevacizumab, Rituxan®/rituximab or Herceptin®/trastuzumab. The combined worldwide market for these three antineoplastic antibodies in 2012 was US$ 21B and the peak sales are projected to reach US$ 25B by 2016. By 2017, all three will loose patent protection in both the US and EU. The effector functions (antibody-dependent cell cytotoxicity, or ADCC) and consequently the clinical efficacy of existing licensed therapeutic antibodies has been found to be strongly inhibited due to nonspecific IgG competing for binding of the therapeutics to FcγRIIIa receptor on natural killer cells, leading to the requirement of significantly higher amount of drug. This problem has been effectively solved by using a production host that does not add a fucose to the glycan structure. Available glycoengineered plants include those with silenced or removed fucosyl transferases, thus the plant-based production provides an attractive opportunity of manufacturing nonfucosylated 'biobetter' molecules with significantly improved ADCC effector function and improved clinical efficacy. Potentially, the ability to use a lower dose of plant-made drug because of its higher efficacy suggests that drug-related adverse events, such as infusion site reactions and other toxicities, could also be minimized. Companies using other manufacturing hosts, including traditional mammalian cells, have also developed nonfucosylating cell lines; therefore, success in the commercial setting will depend on the speed of product development, regulatory treatment of plant-made biosimilars and biobetters, substitutability and reimbursement issues, and other business components rather than on simply the availability of plant production hosts lacking fucose or even potentially lower costs of manufacturing.

3.2 Glycoengineered Plants: A Rapidly Evolving Research Field

A number of research groups have engineered plants to perform "designer" glycosylation. As a first step, galactose residues have been added to protein glycans via expression of proper galactosyl-transferases (Palacpac et al. 1999; Bakker et al. 2001). As a second step, plants lacking plant-specific sugars xylose and α-1,3-linked fucose have been created using RNAi technology (Cox et al. 2006; Schähs et al. 2007; Strasser et al. 2008) and recently, by chemical mutagenesis (Bayer-CropScience). The final missing step, addition of N-acetylneuraminic acid (sialic acid), requires co-expression of 5–6 genes; it has been demonstrated in transient assays (Decker and Reski 2008), but it will take several years before such engineered plants will be available as industrial production hosts. It should be mentioned that such designer glycosylation does not have to be done as a stable genetic transformation. For example, Vezina et al. (2009) demonstrated that transient co-expression of the genes encoding the protein of interest together with the RNAs suppressing fucosyl- and xylosyl-transferases provides for the appropriately modified recombinant proteins.

4 Plant-Based Manufacturing and Downstream Issues

4.1 Current Status of the Industrial Immunoaffinity Separation Process

Purification of pharmaceutical and other high-value proteins is the most important concern for biotechnologists, since commercial downstream processes employed today are extremely complex and costly, accounting for as much as 50–80 % of total manufacturing costs (Roque et al. 2004). Much of that cost is due to the use of highly efficient but very expensive affinity chromatography. In particular, the cost of Protein A-based immunoadsorbent, a standard reagent used for industrial purification of monoclonal antibodies, is so high that the packaged columns have to be re-used up to 50 times, resulting in significant costs due to the necessity for multiple cleanings and re-validation of the columns. An ideal alternative would be a low cost, single-use reagent that performs well and provides a sufficient economic incentive for industrial adoption.

4.2 Biotechnical Solutions to Separation Challenges

Several potential biotechnological solutions have been described in the literature that propose fusing Protein A (or other affinity ligands) to elements capable of polymerizing or to structures of high molecular weight. Such candidates include bacterial S-layer proteins (Vollenkle et al. 2004; Nomellini et al. 2007), bacteriophages (Kushwaha et al. 1994), oleosins recognizing oil bodies (see below), cellulose-binding domains (Cao et al. 2007), and other moieties. Only the process based on oleosin fusions (to Protein A) has been advanced to the stage of a commercially viable platform by SemBioSys Genetics (Canada); however, the company ceased operations recently in spite of their technical innovation. One serious limitation of that technology was the relatively low yield of oleosin fusion molecules (around 1 kg or less per hectare of safflower planting).

Due to their relatively simple macromolecular organization and very high accumulation titers, plant tobamoviruses provide an extremely inexpensive source of protein biopolymers of discrete nanoparticle sizes than can be manufactured rapidly and under very simple conditions. In particular, it has been found that a functional fragment of Protein A (133 aa) can be displayed on the surface of TMV as a carboxy-terminal fusion to the CP via a 15 aa linker. The macromolecular nature of these nanoparticles allowed the design of a simple laboratory protocol for purification of monoclonal antibodies with 50 % recovery yield and >90 % product purity. The extremely dense packing of Protein A on the nanoparticles (over 2100 copies per TMV particle) results in an immunoadsorbent with a binding capacity of 2 g mAb per g nanoparticle. This characteristic, combined with the high level of expression of the nanoparticles (up to 5 g adsorbent per kg of leaf biomass), provides a very inexpensive self-assembling matrix that could meet criteria for a single-use industrial immunoadsorbent for antibody purification (Werner et al. 2006). For industrial application, the described immunoadsorbent still needs to be improved by adding at least one other functionality, so as to allow easy separation of the nanoparticle from the rest of molecules in the extract.

5 Biosafety of Products and Industrial Processes Based on Plant Viral Vectors and Magnifection

5.1 Safety of Plant-Made Pharmaceutical Products

It is generally agreed that, unlike other production organisms used for manufacturing of biopharmaceuticals, plant cells do not harbor pathogens capable of infecting humans. Materials used to grow plants and to cultivate agrobacteria are, or can be selected to be, all animal component-free. In this respect, plant-derived recombinant proteins are inherently safer than materials generated using animal cells. In particular, FDA requires no less than three purification steps in the downstream purification of any animal cell-derived product to assure proper dilution of any remaining pathogen in the purified product. Such requirements are probably not needed in the case of plant-derived products, provided other purity criteria are met. Specific issues potentially impacting the safety of viral vector- and magnifection-derived products are summarized below.

Initially, there were concerns that RNA virus-derived protein identity could be compromised, because of low but measurable frequency of sequence drift in plant viral RNAs during replication. However, experimental data have demonstrated later that foreign sequences propagated in TMV-based RNA vectors show virtually no sequence drift, even during multiple passages in whole plants (Kearney et al. 1993; Turpen 1999); this lack of drift allowed LSBC and subsequently several other groups to successfully perform a number of preclinical and clinical trials with proteins generated using viral vectors based on TMV or other tobamoviruses. An example of the FDA's acceptance of such an expression system is the successful Phase I clinical trial with scFv molecules prepared as individualized vaccines for non-Hodgkin's lymphoma conducted by LSBC under FDA IND (McCormick et al. 2008). Compared to the 'full virus' vectors, replicons under the magnifection process undergo even fewer replication cycles (there is no secondary replication stage due to systemic spread of the vector), thus further lowering any risk related to sequence drift. This position is also accepted by the FDA, reflected in the fact that FDA approved Icon Genetics' IND for a clinical trial of individualized vaccines for non-Hodgkin's lymphoma comprising immunoglobulins expressed using two viral vectors, one based on tobamovirus and the other on PVX. Similarly, FDA approval was gained by Fraunhofer USA for a clinical trial of influenza hemagglutinin vaccine produced in a TMV-based transient expression system.

5.2 Safety of Other Products Derived From Transient Expression

Product requirements for plant-made products other than biopharmaceuticals, such as industrial enzymes, food products and biomaterials, are less stringent than those for biopharmaceuticals and such products derived from transient expression will likely be approved. One interesting approval route in the USA available today is the so-called GRAS (Generally Recognized As Safe) product approval pathway. Such an approval is possible because the inherent safety of the components and individual steps of magnifection-based manufacturing can be documented (expression can be done in edible plants such as beets; both TMV/PVX virus and *Agrobacterium* are part of our daily diet because they are consumed together with raw vegetables; and the product per se can be a part of a GRAS organism or part of our diet as well).

5.3 Process/Environmental Safety in Contained Facilities

The high expression levels achieved in producing most pharmaceutical and veterinary proteins enable the whole production process to be practiced in a contained facility, thus fully mitigating environmental biosafety related risk. Plant-based production entails a protocol similar to existing industrial microbial technologies; it requires, in addition to well-established industrial upstream (greenhouse plant cultivation) and downstream (protein extraction and purification) components, a contained technology facility that includes equipment for vacuum-infiltration of batches of plants and a chamber/greenhouse for short-term subsequent incubation, as well as a bacterial fermentation apparatus. Such a facility provides physical safety barriers necessary to avoid the release of agrobacteria or plants into the open environment as well as to protect operating personnel. For example, in Icon Genetics' pilot plant in Halle, Germany, which is licensed and certified by German authorities for manufacturing specific products, simple, effective, and inexpensive rules of protection of personnel have been successfully implemented.

5.4 Process/Environmental Safety Upon Release into Open Environments

Because of very high expression levels obtainable in transient expression processes, most high-value, low-volume products could be produced in a contained manufacturing facility. However, there are numerous applications where the targets are lower value, high-volume products where open field cultivation would be necessitated to gain full advantage of agricultural economics. Examples of such targets include production of cellulases for cellulosic bioethanol (potential need would be in the tens of thousands of tons/year or even higher), or use of magnifection to modify the agronomic performance of crops. Such applications would require release of genetically engineered agrobacteria into the environment and the prerequisite for any such release would be further studies and measures to assure the biological and environmental safety of the technology. In this regard, it should be noted, that *Agrobacterium* is a ubiquitous soil bacterium that is present in organically rich soils at levels as high as 10^7 bacteria per cm^3 of soil (Krimi et al. 2002), making it an unavoidable part of our diet due to our consumption of raw leafy vegetables, most of which provide natural habitats for agrobacteria. Our current (partially optimized) prototype of the process is already relatively efficient with regard to bacterial inocula needed, such that the treatment of 1 ton of host plant biomass can be achieved with $\sim$10 g of bacteria, or 10^{12} CFUs. At this application rate, the amount of released bacteria per hectare will be less than 0.1 % of agrobacteria already present in the upper 1 cm of soil (i.e., 10^{12} sprayed versus 10^{15} naturally present). Additionally, the *Agrobacterium* intended for release can be disabled in multiple ways to further restrict its potential environmental impact, including its ability to survive, transfect plants or transfer its T-DNA to other bacteria (Gleba and Giritch 2011). At least in the USA, there is a regulatory path for registering GM microorganisms intended for open release and the use of several such organisms has already been approved, including *Agrobacterium radiobacter* strain K1026 used industrially for controlling phytopathogenic agrobacteria (Wozniak et al. 2012).

5.5 Additional Biosafety Features that Could be Introduced Through Genetic Manipulation

The magnifection process could be made still safer through genetic manipulations of the industrial vectors used. In particular, only the viral replicons that are defective and cannot generate a wild-type virus or effectively re-infect other plants will be used in magnifection; in addition to the unneeded functions (systemic spread and infectious virion formation) that have already been eliminated from the TMV vector, some useful functions such as the ability for cell-to-cell spread can also be removed from the replicon. The agrobacteria used as vectors are ubiquitous in the soil and, for magnifection applications, can, and already have partially been made, genetically disabled through various manipulations. 'Disarming' treatments include (a) introduction of multiple auxotrophy and other genetic modifications making such bacteria unable to survive in natural environments, (b) deficiency in conjugative plasmid transfer, and (c) conditional deficiency in transfer of functional genes to a plant. The replicons (and optionally the gene of interest) encoded in the T-DNA contain multiple introns and cannot be expressed in any microbial organism as a result of unintentional horizontal transfer (Gleba et al. 2004, 2005, 2007; Gleba and Giritch 2011).

6 Epilog

In his famous (and infamous) 1997 interview to 'Harvard Business Reviews', Robert Shapiro, then CEO of Monsanto, very eloquently (and passionately) described plant biotechnology and his vision of the future impact of this technology on society. He said: "Sustainability involves the laws of nature—physics, chemistry, and biology—and the recognition that the world is a closed system. What we thought was boundless has limits, and we're beginning to hit them... We don't have 100 years to figure that out; at best, we have decades... In that time frame, I know of only two viable candidates: biotechnology and information technology. I'm treating them as though they're separate, but biotechnology is really a subset of information technology because it is about DNA-encoded information. I offer a prediction: the early twenty-first century is going to see a struggle between information technology and biotechnology on the one hand and environmental degradation on the other. Information technology is going to be our most powerful tool. It will let us miniaturize things, avoid waste, and produce more value without producing and processing more stuff. The substitution of information for stuff is essential to sustainability."

These assessments and prophecies are most interesting in two respects. First, not too surprisingly, many of his predictions seem to be slow to come. The reality is that today, 16 years since that interview (the time frame realistically needed to take a product from early discovery to the market!), the fruits of industrial plant biotechnology comprise only a couple of agronomic traits in three major crops, the world market value of which in 2012 amounted to a mere $15 billion (data released by ISAAA, www.isaaa.org). Second, and more important, during the subsequent one and one half decades, we have witnessed unprecedented accumulation of knowledge in plant molecular biology and biotechnology, a development that is more and more transforming our academic knowledge of the green plant as a molecular machine into an 'actionable intelligence', a collection of powerful biotechnologies. In this regard, perhaps somewhat subjectively, we view transient technologies described in this book as very important new processing tools—the tools that allow us to 'miniaturize things' and 'produce more value without producing and processing more stuff', and perform those manipulations much faster, more flexibly, and cheaper.

The analogy with (much more rapidly developing) information technology is also appropriate here. A transgenic plant expressing one or a few foreign genes is just a one (or two, or three) trick pony, and in that regard it resembles a simple computer of the early 1980s that was able to perform only a couple of general functions such as word processing or arithmetic calculations. Programming such hardware required laborious, large programs enacted through slow, and inefficient inputs and outputs. One the other hand, rapid reprogramming of plant performance on the basis of transient technologies allows for faster, more efficient, and versatile manipulation of a green plant as a production machine. Such biological instructions are analogous to modern computer miniprograms or applications ('Apps'), which are delivered to the plant hardware via *Agrobacterium*. As with computer software, some Apps are better than others at accomplishing their goals. Regardless, with its ability to deliver Apps in a form of DNA, *Agrobacterium* is in this regard very similar to a more recent computer device, a USB memory stick, as it enables the easy transfer of information for the control of a plant's hardware in countless ways to suit each user's needs and preferences (Fig. 2).

Today, the modern nomads travel to the end of the world while staying fully informed about and connected to the rest of the society via cellular telephones and tablet computers. It is not too farfetched to expect that in the foreseeable future, thanks to biotechnologies, we shall be much more liberated also in a material way. Transient expression technologies based on *Agrobacterium* or viral delivery have already demonstrated that, by using very simple processes and very simple equipment (all a result of extensive genetic engineering behind the scene), one can cause green plants to produce food, materials and energy, as well as our own medicine, much more efficiently and in novel ways unavailable to us before. In our experiments, we and others have repeatedly demonstrated the above by producing vaccines, antibodies, therapeutic enzymes and other biopharmaceuticals, as well as biomaterials and biofuel enzymes, and by programming plants for better agronomic performance. We can imagine ourselves in a future where we can go to the middle of nowhere with a rucksack containing small panels of necessary bacterial strains (and a cell phone, of course) and be able to live there comfortably, sustainably, and perhaps even autonomously from the rest of civilization. More importantly, the impact of such technologies on indigenous people who live without the complicated infrastructure of the Western world will be even greater. The business potential of these technologies is also huge: since the advent of genetically engineered plants that were first produced in 1982 and commercialized in 1996, the transient processes described in this book are collectively the only new technology with a truly disruptive potential—the one that allows us to create, control, and extract value in bioindustry in an entirely different way.

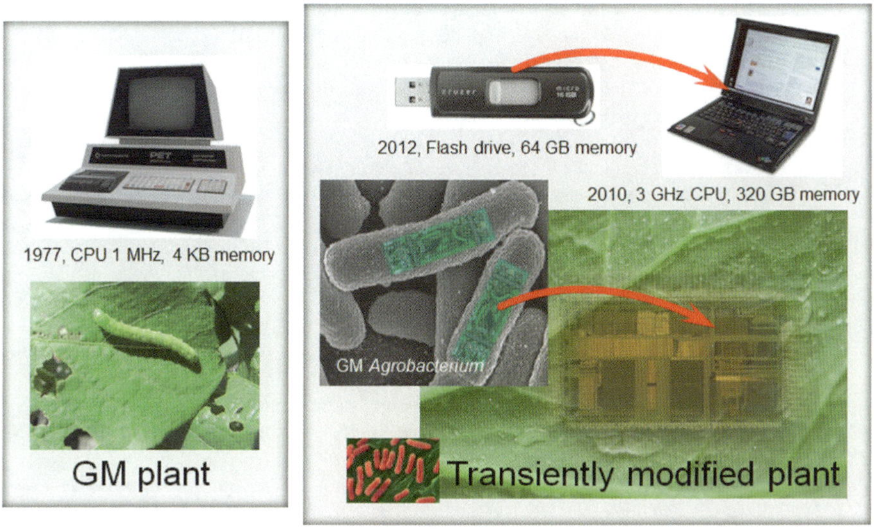

Fig. 2 Stably and transiently modified plants: analogy with computer information processing. Plants modified by stable transgenic methods to express one or a few new traits only perform one to a few new functions. Their genetic code took great effort and time to express reproducibly and reach commercial scale, and once constructed, such hardware was not easily modified. In this sense, transgenic plants resemble computer hardware ca 1980s, which could only run a few basic programs such as word processing, simple spreadsheets or mathematical functions, and whose operating systems were cumbersome to program and slow to access. In sharp contrast, transient expression mediated by plant viruses or *Agrobacterium*-encoded amplicons enables rapid reprogramming of cellular functions to express multiple new traits or produce a wide range of protein products of commercial value. Such nimble genetic programs carried by the transient vectors resemble computer miniprograms or applications ('Apps'), which can perform myriad functions on multiple platforms without altering the hardware. Vectoring via *Agrobacterium* is then analogous to a modern USB flash drive, as both devices can conveniently and rapidly introduce instructions into their intended hardware. Users of modern computers and transient expression technologies thus benefit from the freedom to choose from a wide variety of Apps, input devices, and hardware to suit their needs and preferences in ways that were unimaginable only a decade ago

References

Azhakanandam K, Weissinger SM, Nicholson JS, Qu R, Weissinger AK (2007) Amplicon-plus targeting technology (APTT) for rapid production of a highly unstable vaccine protein in tobacco plants. Plant Mol Biol 63:393–404

Bakker H, Bardor M, Molthoff JW, Gomord V, Elbers I, Stevens LH, Jordi W, Lommen A, Faye L, Lerouge P, Bosch D (2001) Galactose-extended glycans of antibodies produced by transgenic plants. Proc Natl Acad Sci USA 98:2899–2904

Barton KA, Binns AN, Matzke AJ, Chilton MD (1983) Regeneration of intact tobacco plants containing full length copies of genetically engineered T-DNA, and transmission of T-DNA to R1 progeny. Cell 32:1033–1043

Bendandi M, Marillonnet S, Kandzia R, Thieme F, Nickstadt A, Herz S, Frode R, Inoges S, Lopez-Diaz de Cerio A, Soria E, Villanueva H, Vancanneyt G, McCormick A, Tuse D, Lenz

J, Butler-Ransohoff JE, Klimyuk V, Gleba Y (2010) Rapid, high-yield production in plants of individualized idiotype vaccines for non-Hodgkin's lymphoma. Ann Oncol 21:2420–2427

Caddick MX, Greenland AJ, Jepson I, Krause KP, Qu N, Riddell KV, Salter MG, Schuch W, Sonnewald U, Tomsett AB (1998) An ethanol inducible gene switch for plants used to manipulate carbon metabolism. Nat Biotechnol 16:177–180

Canizares MC, Lomonossoff GP, Nicholson L (2005a) Development of cowpea mosaic virus-based vectors for the production of vaccines in plants. Expert Rev Vaccines 4:687–697

Canizares MC, Nicholson L, Lomonossoff GP (2005b) Use of viral vectors for vaccine production in plants. Immunol Cell Biol 83:263–270

Cao Y, Zhang Q, Wang C, Zhu Y, Bai G (2007) Preparation of novel immunomagnetic cellulose microspheres via cellulose binding domain-protein A linkage and its use for the isolation of interferon alpha-2b. J Chromatogr A 1149:228–235

Chen Q, He J, Phoolcharoen W, Mason HS (2011) Geminiviral vectors based on bean yellow dwarf virus for production of vaccine antigens and monoclonal antibodies in plants. Hum Vaccin 7:331–338

Chichester JA, Musiychuk K, de la Rosa P, Horsey A, Stevenson N, Ugulava N, Rabindran S, Palmer GA, Mett V, Yusibov V (2007) Immunogenicity of a subunit vaccine against *Bacillus anthracis*. Vaccine 25:3111–3114

Chichester JA, Manceva SD, Rhee A, Coffin MV, Musiychuk K, Mett V, Shamloul M, Norikane J, Streatfield SJ, Yusibov V (2013) A plant-produced protective antigen vaccine confers protection in rabbits against a lethal aerosolized challenge with *Bacillus anthracis* Ames spores. Hum Vaccin Immunother 9:544–552

Citovsky V, Kozlovsky SV, Lacroix B, Zaltsman A, Dafny-Yelin M, Vyas S, Tovkach A, Tzfira T (2007) Biological systems of the host cell involved in *Agrobacterium* infection. Cell Microbiol 9:9–20

Corrado G, Karali M (2009) Inducible gene expression systems and plant biotechnology. Biotechnol Adv 27:733–743

Cox KM, Sterling JD, Regan JT, Gasdaska JR, Frantz KK, Peele CG, Black A, Passmore D, Moldovan-Loomis C, Srinivasan M, Cuison S, Cardarelli PM, Dickey LF (2006) Glycan optimization of a human monoclonal antibody in the aquatic plant *Lemna minor*. Nat Biotechnol 24:1591–1597

Daniell H, Singh ND, Mason H, Streatfield SJ (2009) Plant-made vaccine antigens and biopharmaceuticals. Trends Plant Sci 14:669–679

Dawson WO, Beck DL, Knorr DA, Grantham GL (1986) cDNA cloning of the complete genome of tobacco mosaic virus and production of infectious transcripts. Proc Natl Acad Sci USA 83:1832–1836

Decker EL, Reski R (2008) Current achievements in the production of complex biopharmaceuticals with moss bioreactors. Bioprocess Biosyst Eng 31:3–9

Dohi K, Nishikiori M, Tamai A, Ishikawa M, Meshi T, Mori M (2006) Inducible virus-mediated expression of a foreign protein in suspension-cultured plant cells. Arch Virol 151:1075–1084

Donson J, Kearney CM, Hilf ME, Dawson WO (1991) Systemic expression of a bacterial gene by a tobacco mosaic virus-based vector. Proc Natl Acad Sci USA 88:7204–7208

Dorokhov YL, Sheveleva AA, Frolova OY, Komarova TV, Zvereva AS, Ivanov PA, Atabekov JG (2007) Superexpression of tuberculosis antigens in plant leaves. Tuberculosis (Edinb) 87:218–224

Engler C, Kandzia R, Marillonnet S (2008) A one pot, one step, precision cloning method with high throughput capability. PLoS ONE 3:e3647

Engler C, Gruetzner R, Kandzia R, Marillonnet S (2009) Golden gate shuffling: a one-pot DNA shuffling method based on type IIs restriction enzymes. PLoS ONE 4:e5553

Evans DJ (2009) Exploitation of plant and archaeal viruses in bionanotechnology. Biochem Soc Trans 37:665–670

Fraley RT, Rogers SG, Horsch RB, Sanders PR, Flick JS, Adams SP, Bittner ML, Brand LA, Fink CL, Fry JS, Galluppi GR, Goldberg SB, Hoffmann NL, Woo SC (1983) Expression of bacterial genes in plant cells. Proc Natl Acad Sci USA 80:4803–4807

Gatz C (1997) Chemical control of gene expression. Annu Rev Plant Physiol Plant Mol Biol 48:89–108

Gelvin SB (2008) *Agrobacterium*-mediated DNA transfer, and then some. Nat Biotechnol 26:998–1000

Gelvin SB (2009) *Agrobacterium* in the genomics age. Plant Physiol 150:1665–1676

Gelvin SB (2010) Plant proteins involved in *Agrobacterium*-mediated genetic transformation. Annu Rev Phytopathol 48:45–68

Gelvin SB (2012) Traversing the cell: *Agrobacterium* T-DNA's journey to the host genome. Front Plant Sci 3:52

Gils M, Kandzia R, Marillonnet S, Klimyuk V, Gleba Y (2005) High-yield production of authentic human growth hormone using a plant virus-based expression system. Plant Biotechnol J 3:613–620

Giritch A, Marillonnet S, Engler C, van Eldik G, Botterman J, Klimyuk V, Gleba Y (2006) Rapid high-yield expression of full-size IgG antibodies in plants coinfected with noncompeting viral vectors. Proc Natl Acad Sci USA 103:14701–14706

Giritch A, Symonenko Y, Hahn S, Tiede D, Shvarts A, Roemer P, Gleba Y (2012) Process of Transfecting Plants. Patent WO/2012/019660

Gleba Y (2013) Technology progress in PMP research. In: The Fifth International Conference on Plant-Based Vaccines, Antibodies and Biologics, Verona, Italy, 5-7 June 2013 (abstract)

Gleba YY, Giritch A (2011) Plant Viral Vectors for Protein Expression. In: Caranta C, Aranda MA, Tepfer M, Lopez-Moya JJ (eds) Recent Advances in Plant Virology. Caister Academic Press, Norfolk, pp 387–412

Gleba Y, Marillonnet S, Klimyuk V (2004) Engineering viral expression vectors for plants: the 'full virus' and the 'deconstructed virus' strategies. Curr Opin Plant Biol 7:182–188

Gleba Y, Klimyuk V, Marillonnet S (2005) Magnifection-a new platform for expressing recombinant vaccines in plants. Vaccine 23:2042–2048

Gleba Y, Klimyuk V, Marillonnet S (2007) Viral vectors for the expression of proteins in plants. Curr Opin Biotechnol 18:134–141

Gleba Y, Marillonnet S, Klimyuk V (2008) Plant virus vectors (gene expression systems). In: van Regenmortel MHV, Mahy BWJ (eds) Encyclopedia of Virology. Elsevier Academic Press, Oxford, pp 229–237

Golovkin M, Spitsin S, Andrianov V, Smirnov Y, Xiao Y, Pogrebnyak N, Markley K, Brodzik R, Gleba Y, Isaacs SN, Koprowski H (2007) Smallpox subunit vaccine produced in Planta confers protection in mice. Proc Natl Acad Sci USA 104:6864–6869

Grimsley N, Hohn B, Hohn T, Walden R (1986) "Agroinfection," an alternative route for viral infection of plants by using the Ti plasmid. Proc Natl Acad Sci USA 83:3282–3286

Hahn S, Giritch A, Gleba Y (2013) Production, storage and use of cell wall-degrading enzymes. Patent EP 2(584):042

Hamorsky KT, Grooms-Williams TW, Husk AS, Bennett LJ, Palmer KE, Matoba N (2013a) Efficient single tobamoviral vector-based bioproduction of broadly neutralizing anti-HIV-1 monoclonal antibody VRC01 in *Nicotiana benthamiana* plants and utility of VRC01 in combination microbicides. Antimicrob Agents Chemother 57:2076–2086

Hamorsky KT, Kouokam JC, Bennett LJ, Baldauf KJ, Kajiura H, Fujiyama K, Matoba N (2013b) Rapid and scalable plant-based production of a cholera toxin B subunit variant to aid in mass vaccination against cholera outbreaks. PLoS Negl Trop Dis 7:e2046

Hiatt A, Pauly M (2006) Monoclonal antibodies from plants: a new speed record. Proc Natl Acad Sci USA 103:14645–14646

Hiatt A, Cafferkey R, Bowdish K (1989) Production of antibodies in transgenic plants. Nature 342:76–78

Huang Z, Santi L, LePore K, Kilbourne J, Arntzen CJ, Mason HS (2006) Rapid, high-level production of hepatitis B core antigen in plant leaf and its immunogenicity in mice. Vaccine 24:2506–2513

Huang Z, Chen Q, Hjelm B, Arntzen C, Mason H (2009) A DNA replicon system for rapid high-level production of virus-like particles in plants. Biotechnol Bioeng 103:706–714

Huang Z, Phoolcharoen W, Lai H, Piensook K, Cardineau G, Zeitlin L, Whaley KJ, Arntzen CJ, Mason HS, Chen Q (2010) High-level rapid production of full-size monoclonal antibodies in plants by a single-vector DNA replicon system. Biotechnol Bioeng 106:9–17

Hull AK, Yusibov V, Mett V (2005) Inducible expression in plants by virus-mediated transgene activation. Transgenic Res 14:407–416

Karg SR, Kallio PT (2009) The production of biopharmaceuticals in plant systems. Biotechnol Adv 27:879–894

Kearney CM, Donson J, Jones GE, Dawson WO (1993) Low level of genetic drift in foreign sequences replicating in an RNA virus in plants. Virology 192:11–17

Klimyuk V, Pogue G, Herz S, Butler J, Haydon H (2012) Production of recombinant antigens and antibodies in *Nicotiana benthamiana* using 'Magnifection' technology: GMP-compliant facilities for small- and large-scale manufacturing. Curr Top Microbiol Immunol. doi:10.1007/82_2012_212

Kohl T, Hitzeroth II, Stewart D, Varsani A, Govan VA, Christensen ND, Williamson AL, Rybicki EP (2006) Plant-produced cottontail rabbit papillomavirus L1 protein protects against tumor challenge: a proof-of-concept study. Clin Vaccine Immunol 13:845–853

Komarova TV, Baschieri S, Donini M, Marusic C, Benvenuto E, Dorokhov YL (2010) Transient expression systems for plant-derived biopharmaceuticals. Expert Rev Vaccines 9:859–876

Krimi Z, Petit A, Mougel C, Dessaux Y, Nesme X (2002) Seasonal fluctuations and long-term persistence of pathogenic populations of *Agrobacterium* spp. in soils. Appl Environ Microbiol 68:3358–65

Kumagai MH, Turpen TH, Weinzettl N, della-Cioppa G, Turpen AM, Donson J, Hilf ME, Grantham GL, Dawson WO, Chow TP, Piatak Jr M, Grill LK (1993) Rapid, high-level expression of biologically active alpha-trichosanthin in transfected plants by an RNA viral vector. Proc Natl Acad Sci USA 90:427–430

Kumagai MH, Donson J, della-Cioppa G, Harvey D, Hanley K, Grill LK (1995) Cytoplasmic inhibition of carotenoid biosynthesis with virus-derived RNA. Proc Natl Acad Sci USA 92:1679–1683

Kumagai MH, Donson J, della-Cioppa G, Grill LK (2000) Rapid, high-level expression of glycosylated rice alpha-amylase in transfected plants by an RNA viral vector. Gene 245:169–174

Kushwaha A, Chowdhury PS, Arora K, Abrol S, Chaudhary VK (1994) Construction and characterization of M13 bacteriophages displaying functional IgG-binding domains of staphylococcal protein A. Gene 151:45–51

Lico C, Chen Q, Santi L (2008) Viral vectors for production of recombinant proteins in plants. J Cell Physiol 216:366–377

Lindbo JA (2007a) TRBO: a high-efficiency tobacco mosaic virus RNA-based overexpression vector. Plant Physiol 145:1232–1240

Lindbo JA (2007b) High-efficiency protein expression in plants from agroinfection-compatible Tobacco mosaic virus expression vectors. BMC Biotechnol 7:52

Ma JK, Hiatt A, Hein M, Vine ND, Wang F, Stabila P, van Dolleweerd C, Mostov K, Lehner T (1995) Generation and assembly of secretory antibodies in plants. Science 268:716–719

Maclean J, Koekemoer M, Olivier AJ, Stewart D, Hitzeroth II, Rademacher T, Fischer R, Williamson AL, Rybicki EP (2007) Optimization of human papillomavirus type 16 (HPV-16) L1 expression in plants: comparison of the suitability of different HPV-16 L1 gene variants and different cell-compartment localization. J Gen Virol 88:1460–1469

Magori S, Citovsky V (2011) Epigenetic control of *Agrobacterium* T-DNA integration. Biochim Biophys Acta 1809:388–394

Mallory AC, Parks G, Endres MW, Baulcombe D, Bowman LH, Pruss GJ, Vance VB (2002) The amplicon-plus system for high-level expression of transgenes in plants. Nat Biotechnol 20:622–625

Marillonnet S, Giritch A, Gils M, Kandzia R, Klimyuk V, Gleba Y (2004) *In planta* engineering of viral RNA replicons: efficient assembly by recombination of DNA modules delivered by *Agrobacterium*. Proc Natl Acad Sci USA 101:6852–6857

Marillonnet S, Thoeringer C, Kandzia R, Klimyuk V, Gleba Y (2005) Systemic *Agrobacterium tumefaciens*-mediated transfection of viral replicons for efficient transient expression in plants. Nat Biotechnol 23:718–723

Massa S, Franconi R, Brandi R, Muller A, Mett V, Yusibov V, Venuti A (2007) Anti-cancer activity of plant-produced HPV16 E7 vaccine. Vaccine 25:3018–3021

McCormick AA (2011) Tobacco derived cancer vaccines for non-Hodgkin's lymphoma: perspectives and progress. Hum Vaccin 7:305–312

McCormick AA, Palmer KE (2008) Genetically engineered Tobacco mosaic virus as nanoparticle vaccines. Expert Rev Vaccines 7:33–41

McCormick AA, Kumagai MH, Hanley K, Turpen TH, Hakim I, Grill LK, Tuse D, Levy S, Levy R (1999) Rapid production of specific vaccines for lymphoma by expression of the tumor-derived single-chain Fv epitopes in tobacco plants. Proc Natl Acad Sci USA 96:703–708

McCormick AA, Reinl SJ, Cameron TI, Vojdani F, Fronefield M, Levy R, Tuse D (2003) Individualized human scFv vaccines produced in plants: humoral anti-idiotype responses in vaccinated mice confirm relevance to the tumor Ig. J Immunol Methods 278:95–104

McCormick AA, Reddy S, Reinl SJ, Cameron TI, Czerwinkski DK, Vojdani F, Hanley KM, Garger SJ, White EL, Novak J, Barrett J, Holtz RB, Tuse D, Levy R (2008) Plant-produced idiotype vaccines for the treatment of non-Hodgkin's lymphoma: safety and immunogenicity in a phase I clinical study. Proc Natl Acad Sci USA 105:10131–10136

Mett V, Lyons J, Musiychuk K, Chichester JA, Brasil T, Couch R, Sherwood R, Palmer GA, Streatfield SJ, Yusibov V (2007) A plant-produced plague vaccine candidate confers protection to monkeys. Vaccine 25:3014–3017

Mett V, Farrance CE, Green BJ, Yusibov V (2008) Plants as biofactories. Biologicals 36:354–358

Meyers A, Chakauya E, Shephard E, Tanzer FL, Maclean J, Lynch A, Williamson AL, Rybicki EP (2008) Expression of HIV-1 antigens in plants as potential subunit vaccines. BMC Biotechnol 8:53

Mor TS, Moon YS, Palmer KE, Mason HS (2003) Geminivirus vectors for high-level expression of foreign proteins in plant cells. Biotechnol Bioeng 81:430–437

Mortimer E, Maclean JM, Mbewana S, Buys A, Williamson AL, Hitzeroth II, Rybicki EP (2012) Setting up a platform for plant-based influenza virus vaccine production in South Africa. BMC Biotechnol 12:14

Nomellini JF, Duncan G, Dorocicz IR, Smit J (2007) S-layer-mediated display of the immunoglobulin G-binding domain of streptococcal protein G on the surface of *Caulobacter crescentus*: development of an immunoactive reagent. Appl Environ Microbiol 73:3245–3253

O'Keefe BR, Vojdani F, Buffa V, Shattock RJ, Montefiori DC, Bakke J, Mirsalis J, d'Andrea AL, Hume SD, Bratcher B, Saucedo CJ, McMahon JB, Pogue GP, Palmer KE (2009) Scaleable manufacture of HIV-1 entry inhibitor griffithsin and validation of its safety and efficacy as a topical microbicide component. Proc Natl Acad Sci USA 106:6099–6104

Padidam M (2003) Chemically regulated gene expression in plants. Curr Opin Plant Biol 6:169–177

Padidam M, Gore M, Lu DL, Smirnova O (2003) Chemical-inducible, ecdysone receptor-based gene expression system for plants. Transgenic Res 12:101–109

Palacpac NQ, Yoshida S, Sakai H, Kimura Y, Fujiyama K, Yoshida T, Seki T (1999) Stable expression of human beta1,4-galactosyltransferase in plant cells modifies N-linked glycosylation patterns. Proc Natl Acad Sci U S A 96:4692–4697

Pelosi A, Shepherd R, Walmsley AM (2012) Delivery of plant-made vaccines and therapeutics. Biotechnol Adv 30:440–448

Pogue GP, Vojdani F, Palmer KE, Hiatt E, Hume S, Phelps J, Long L, Bohorova N, Kim D, Pauly M, Velasco J, Whaley K, Zeitlin L, Garger SJ, White E, Bai Y, Haydon H, Bratcher B (2010) Production of pharmaceutical-grade recombinant aprotinin and a monoclonal antibody product using plant-based transient expression systems. Plant Biotechnol J 8:638–654

Rabindran S, Dawson WO (2001) Assessment of recombinants that arise from the use of a TMV-based transient expression vector. Virology 284:182–189

Reavy B, Bagirova S, Chichkova NV, Fedoseeva SV, Kim SH, Vartapetian AB, Taliansky ME (2007) Caspase-resistant VirD2 protein provides enhanced gene delivery and expression in plants. Plant Cell Rep 26:1215–1219

Regnard GL, Halley-Stott RP, Tanzer FL, Hitzeroth II, Rybicki EP (2010) High level protein expression in plants through the use of a novel autonomously replicating geminivirus shuttle vector. Plant Biotechnol J 8:38–46

Roque AC, Lowe CR, Taipa MA (2004) Antibodies and genetically engineered related molecules: production and purification. Biotechnol Prog 20:639–654

Royston E, Ghosh A, Kofinas P, Harris MT, Culver JN (2008) Self-assembly of virus-structured high surface area nanomaterials and their application as battery electrodes. Langmuir 24:906–912

Rybicki EP (2009) Plant-produced vaccines: promise and reality. Drug Discov Today 14:16–24

Ryu CM, Anand A, Kang L, Mysore KS (2004) Agrodrench: a novel and effective agroinoculation method for virus-induced gene silencing in roots and diverse *Solanaceous* species. Plant J 40:322–331

Sainsbury F, Lomonossoff GP (2008) Extremely high-level and rapid transient protein production in plants without the use of viral replication. Plant Physiol 148:1212–1218

Sainsbury F, Liu L, Lomonossoff GP (2009) Cowpea mosaic virus-based systems for the expression of antigens and antibodies in plants. Methods Mol Biol 483:25–39

Sainsbury F, Sack M, Stadlmann J, Quendler H, Fischer R, Lomonossoff GP (2010) Rapid transient production in plants by replicating and non-replicating vectors yields high quality functional anti-HIV antibody. PLoS One 5:e13976

Santi L, Giritch A, Roy CJ, Marillonnet S, Klimyuk V, Gleba Y, Webb R, Arntzen CJ, Mason HS (2006) Protection conferred by recombinant *Yersinia pestis* antigens produced by a rapid and highly scalable plant expression system. Proc Natl Acad Sci USA 103:861–866

Santi L, Batchelor L, Huang Z, Hjelm B, Kilbourne J, Arntzen CJ, Chen Q, Mason HS (2008) An efficient plant viral expression system generating orally immunogenic Norwalk virus-like particles. Vaccine 26:1846–1854

Saunders K, Lomonossoff GP (2013) Exploiting plant virus-derived components to achieve in planta expression and for templates for synthetic biology applications. New Phytol. doi:10.1111/nph.12204

Schähs M, Strasser R, Stadlmann J, Kunert R, Rademacher T, Steinkellner H (2007) Production of a monoclonal antibody in plants with a humanized N-glycosylation pattern. Plant Biotechnol J 5:657–663

Shivprasad S, Pogue GP, Lewandowski DJ, Hidalgo J, Donson J, Grill LK, Dawson WO (1999) Heterologous sequences greatly affect foreign gene expression in tobacco mosaic virus-based vectors. Virology 255:312–323

Shoji Y, Chichester JA, Bi H, Musiychuk K, de la Rosa P, Goldschmidt L, Horsey A, Ugulava N, Palmer GA, Mett V, Yusibov V (2008) Plant-expressed HA as a seasonal influenza vaccine candidate. Vaccine 26:2930–2934

Shoji Y, Jones RM, Mett V, Chichester JA, Musiychuk K, Sun X, Tumpey TM, Green BJ, Shamloul M, Norikane J, Bi H, Hartman CE, Bottone C, Stewart M, Streatfield SJ, Yusibov V (2013) A plant-produced H1N1 trimeric hemagglutinin protects mice from a lethal influenza virus challenge. Hum Vaccin Immunother 9:553–560

Smith ML, Fitzmaurice WP, Turpen TH, Palmer KE (2009) Display of peptides on the surface of tobacco mosaic virus particles. Curr Top Microbiol Immunol 332:13–31

Steinmetz NF, Calder G, Lomonossoff GP, Evans DJ (2006) Plant viral capsids as nanobuilding blocks: construction of arrays on solid supports. Langmuir 22:10032–10037

Strasser R, Stadlmann J, Schähs M, Stiegler G, Quendler H, Mach L, Glossl J, Weterings K, Pabst M, Steinkellner H (2008) Generation of glyco-engineered *Nicotiana benthamiana* for the production of monoclonal antibodies with a homogeneous human-like N-glycan structure. Plant Biotechnol J 6:392–402

Terpe K (2006) Overview of bacterial expression systems for heterologous protein production: from molecular and biochemical fundamentals to commercial systems. Appl Microbiol Biotechnol 72:211–222
Thieme F, Engler C, Kandzia R, Marillonnet S (2011) Quick and clean cloning: a ligation-independent cloning strategy for selective cloning of specific PCR products from non-specific mixes. PLoS One 6:e20556
Thuenemann EC, Meyers AE, Verwey J, Rybicki EP, Lomonossoff GP (2013) A method for rapid production of heteromultimeric protein complexes in plants: assembly of protective bluetongue virus-like particles. Plant Biotechnol J. doi:10.1111/pbi.12076
Tremblay A, Beauchemin C, Seguin A, Laliberte JF (2007) Reactivation of an integrated disabled viral vector using a Cre-loxP recombination system in *Arabidopsis thaliana*. Transgenic Res 16:213–222
Turpen TH (1999) Tobacco mosaic virus and the virescence of biotechnology. Philos Trans R Soc Lond B Biol Sci 354:665–673
Turpen TH, Turpen AM, Weinzettl N, Kumagai MH, Dawson WO (1993) Transfection of whole plants from wounds inoculated with *Agrobacterium tumefaciens* containing cDNA of tobacco mosaic virus. J Virol Methods 42:227–239
Turpen TH, Reinl SJ, Charoenvit Y, Hoffman SL, Fallarme V, Grill LK (1995) Malarial epitopes expressed on the surface of recombinant tobacco mosaic virus. Biotechnology (NY) 13:53–57
Tusé D (2011) Safety of plant-made pharmaceuticals: product development and regulatory considerations based on case studies of two autologous human cancer vaccines. Hum Vaccin 7:322–330
Usha R, Rohll JB, Spall VE, Shanks M, Maule AJ, Johnson JE, Lomonossoff GP (1993) Expression of an animal virus antigenic site on the surface of a plant virus particle. Virology 197:366–374
Vancanneyt G, Dubald M, Schroder W, Peters J, Botterman J (2009) A case study for plant-made pharmaceuticals comparing different plant expression and production systems. Methods Mol Biol 483:209–221
Vaquero C, Sack M, Chandler J, Drossard J, Schuster F, Monecke M, Schillberg S, Fischer R (1999) Transient expression of a tumor-specific single-chain fragment and a chimeric antibody in tobacco leaves. Proc Natl Acad Sci USA 96:11128–11133
Verch T, Yusibov V, Koprowski H (1998) Expression and assembly of a full-length monoclonal antibody in plants using a plant virus vector. J Immunol Methods 220:69–75
Vezina LP, Faye L, Lerouge P, D'Aoust MA, Marquet-Blouin E, Burel C, Lavoie PO, Bardor M, Gomord V (2009) Transient co-expression for fast and high-yield production of antibodies with human-like N-glycans in plants. Plant Biotechnol J 7:442–455
Voinnet O, Rivas S, Mestre P, Baulcombe D (2003) An enhanced transient expression system in plants based on suppression of gene silencing by the p19 protein of tomato bushy stunt virus. Plant J 33:949–956
Vollenkle C, Weigert S, Ilk N, Egelseer E, Weber V, Loth F, Falkenhagen D, Sleytr UB, Sara M (2004) Construction of a functional S-layer fusion protein comprising an immunoglobulin G-binding domain for development of specific adsorbents for extracorporeal blood purification. Appl Environ Microbiol 70:1514–1521
Weber E, Engler C, Gruetzner R, Werner S, Marillonnet S (2011) A modular cloning system for standardized assembly of multigene constructs. PLoS One 6:e16765
Weber W, Fussenegger M (2007) Inducible product gene expression technology tailored to bioprocess engineering. Curr Opin Biotechnol 18:399–410
Webster DE, Wang L, Mulcair M, Ma C, Santi L, Mason HS, Wesselingh SL, Coppel RL (2009) Production and characterization of an orally immunogenic *Plasmodium* antigen in plants using a virus-based expression system. Plant Biotechnol J 7:846–855
Werner S, Marillonnet S, Hause G, Klimyuk V, Gleba Y (2006) Immunoabsorbent nanoparticles based on a tobamovirus displaying protein A. Proc Natl Acad Sci USA 103:17678–17683

Werner S, Breus O, Symonenko Y, Marillonnet S, Gleba Y (2011) High-level recombinant protein expression in transgenic plants by using a double-inducible viral vector. Proc Natl Acad Sci USA 108:14061–14066

Whaley KJ, Hiatt A, Zeitlin L (2011) Emerging antibody products and *Nicotiana* manufacturing. Hum Vaccin 7:349–356

Willmitzer L, De Beuckeleer M, Lemmers M, Van Montagu M, Schell J (1980) DNA from Ti plasmid present in nucleus and absent from plastids of crown gall plant cells. Nature 287:359–361

Wozniak CA, Waggoner AF, Reilly S (2012) An introduction to agricultural biotechnology regulation in the US. In: Wozniak CA, McHughen A (eds) Regulation of agricultural biotechnology: the United States and Canada. Springer, Dordrecht, pp 1–15

Yadav NS, Postle K, Saiki RK, Thomashow MF, Chilton MD (1980) T-DNA of a crown gall teratoma is covalently joined to host plant DNA. Nature 287:458–461

Yang L, Wang H, Liu J, Li L, Fan Y, Wang X, Song Y, Sun S, Wang L, Zhu X, Wang X (2008) A simple and effective system for foreign gene expression in plants via root absorption of agrobacterial suspension. J Biotechnol 134:320–324

Young M, Willits D, Uchida M, Douglas T (2008) Plant viruses as biotemplates for materials and their use in nanotechnology. Annu Rev Phytopathol 46:361–384

Yusibov V, Rabindran S (2008) Recent progress in the development of plant derived vaccines. Expert Rev Vaccines 7:1173–1183

Zhang X, Mason H (2006) Bean Yellow Dwarf Virus replicons for high-level transgene expression in transgenic plants and cell cultures. Biotechnol Bioeng 93:271–279

Zuo J, Niu QW, Chua NH (2000) Technical advance: an estrogen receptor-based transactivator XVE mediates highly inducible gene expression in transgenic plants. Plant J 24:265–273

Index

Current Topics in Microbiology and Immunology 375: 193–194
DOI: 10.1007/978-3-642-40829-8

Printed by Printforce, the Netherlands